ADVANCES IN PLANT MICROBIOME RESEARCH FOR CLIMATE-RESILIENT AGRICULTURE

Toward Sustainable Farming

ADVANCES IN PLANT MICROBIOME RESEARCH FOR CLIMATE-RESILIENT AGRICULTURE

Toward Sustainable Farming

Edited by

Ashwani Kumar, PhD

APPLE ACADEMIC PRESS

First edition published 2025

Apple Academic Press Inc.
1265 Goldenrod Circle, NE,
Palm Bay, FL 32905 USA

760 Laurentian Drive, Unit 19,
Burlington, ON L7N 0A4, CANADA

CRC Press
2385 NW Executive Center Drive,
Suite 320, Boca Raton FL 33431

4 Park Square, Milton Park,
Abingdon, Oxon, OX14 4RN UK

Library and Archives Canada Cataloguing in Publication

Title: Advances in plant microbiome research for climate-resilient agriculture : toward sustainable farming / edited by Ashwani Kumar, PhD.
Names: Kumar, Ashwani (Lecturer in botany), editor.
Description: First edition. | Includes bibliographical references and index.
Identifiers: Canadiana (print) 20240341260 | Canadiana (ebook) 2024034135X | ISBN 9781774917640 (hardcover) | ISBN 9781774917657 (softcover) | ISBN 9781003501893 (ebook)
Subjects: LCSH: Plant-microbe relationships—Research. | LCSH: Microbiomes—Research. | LCSH: Crops—Effect of stress on. | LCSH: Sustainable agriculture. | LCSH: Crops and climate. | LCSH: Agriculture—Environmental aspects.
Classification: LCC QR351 .A38 2024 | DDC 579/.178—dc23

Library of Congress Cataloging-in-Publication Data

CIP data on file with US Library of Congress

ISBN: 978-1-77491-764-0 (hbk)
ISBN: 978-1-77491-765-7 (pbk)
ISBN: 978-1-00350-189-3 (ebk)

About the Editor

Ashwani Kumar, PhD
Associate Professor, Metagenomics and Secretomics Research Laboratory, Department of Botany, University of Allahabad (A Central University), Prayagraj, Uttar Pradesh, India;
Metagenomics and Secretomics Research Laboratory, Department of Botany, Dr. Harisingh Gour Vishwavidyalaya (A Central University), Sagar, Madhya Pradesh, India

Ashwani Kumar (PhD, IIT Delhi) FISEB, FNIE, YAI, FLS is presently an Associate Professor in the Department of Botany at the University of Allahabad (A Central University), Prayagraj (UP), India. Before joining this position, he worked at Dr. Harisingh Gour Vishwavidyalaya (A Central University), Sagar, India for more than nine years. His research focused on plant–microbe interactions, metagenomics, lignocellulosic biofuels, and bioremediation. He earned his PhD in bioenergy from the Indian Institute of Technology, Delhi, India, and served as a postdoctoral fellow (2011–2013) at the Durban University of Technology and Rhodes University in Grahamstown, South Africa. He earned several awards, including the GSEP Fellowship at the University of Saskatchewan in Canada (2009–2010), the Claude Leon Fellowship (2012–2013), and the NRF Fellowship (2012–2013) at Rhodes University in Grahamstown, South Africa. In 2019, he was awarded the GIAN Program, MHRD, Government of India, Shastri Indo-Canada Mobility Program Fellowship at Concordia University in Montreal, Canada. He is a life member of the BRSI, the NIE, and the ISEB in India, and he was elected as a fellow of the Linnean Society of London, the International Society of Environmental Botanists, and the National Institute of Ecology. He published over 80 articles (Google Scholar citations: 7500; i-10:80, h-index: 40; Scopus citations: 4750), 30 book chapters, and edited one book. He is serving as an editorial board member of several prestigious journals. He has been featured in the top 2% of the World Scientists list four times (2020-2023).

Contents

Contributors

Muhammad Siddique Afridi
Department of Plant Pathology, Federal University of Lavras, Lavras, MG, Brazil

Baber Ali
Department of Plant Sciences, Quaid-i-Azam University, Islamabad, Pakistan

Himanshu Arora
Indian Institute of Technology, CRDT, Hauz Khas, Delhi, India

Kanika Chowdhary
Indian Institute of Technology, CRDT, Hauz Khas, Delhi, India

Fatih Çiğ
Department of Field Crops, Faculty of Agriculture, Siirt University, Siirt, Turkey

Poonam Dehariya
Lab of Mushroom Biology, Department of Botany, Dr. Harisingh Gour University, Sagar, Madhya Pradesh, India

Anamika Dubey
Metagenomics and Secretomics Research Laboratory, Department of Botany, Dr. Harisingh Gour Vishwavidyalaya (A Central University), Sagar, Madhya Pradesh, India

Murat Erman
Department of Field Crops, Faculty of Agriculture, Bursa Uludağ University, Bursa, Turkey.

Eeman Fatima
Department of Botany, Government College University, Lahore, Pakistan

Aqsa Hafeez
Department of Plant Sciences, Quaid-i-Azam University, Islamabad, Pakistan

Muhammad Aamir Iqbal
Department of Agronomy, Faculty of Agriculture, University of Poonch Rawalakot, Rawalakot, AJK, Pakistan

Mohammad Sohidul Islam
Department of Agronomy, Hajee Mohammad Danesh Science and Technology University, Dinajpur, Bangladesh

Muhammad Ammar Javed
Institute of Industrial Biotechnology, Government College University, Lahore, Pakistan

Talha Javed
College of Agriculture, Fujian Agriculture and Forestry University, Fuzhou, China

Mohammad Latif Khan
Metagenomics and Secretomics Research Laboratory, Department of Botany, Dr. Harisingh Gour Vishwavidyalaya (A Central University), Sagar, Madhya Pradesh, India

Ashwani Kumar
Metagenomics and Secretomics Research Laboratory, Department of Botany, University of Allahabad (A Central University), Prayagraj, Uttar Pradesh, India; Metagenomics and Secretomics Research Laboratory, Department of Botany, Dr. Harisingh Gour Vishwavidyalaya (A Central University), Sagar, Madhya Pradesh, India

Akanksha Kumari
Lab of Mushroom Biology, Department of Botany, Dr. Harisingh Gour University, Sagar, Madhya Pradesh, India

Arpna Kumari
Academy of Biology and Biotechnology, Southern Federal University, Rostov-on-Don, Russia

Javeed Lone
Department of Field Crops, Faculty of Agriculture, Siirt University, Siirt, Turkey

Muneer Ahmad Malla
Metagenomics and Secretomics Research Laboratory, Department of Botany, Dr. Harisingh Gour Vishwavidyalaya (A Central University), Sagar, Madhya Pradesh, India; Department of Zoology, Dr. Harisingh Gour University (A Central University), Sagar, Madhya Pradesh, India

Flavio Henrique Vasconcelos De Medeiros
Department of Plant Pathology, Federal University of Lavras, Lavras, MG, Brazil

Sheo Mohan Prasad
Ranjan Plant Physiology and Biochemistry Laboratory, Department of Botany, University of Allahabad, Prayagraj, Uttar Pradesh, India

Aman Raj
Metagenomics and Secretomics Research Laboratory, Department of Botany, Dr. Harisingh Gour Vishwavidyalaya (A Central University), Sagar, Madhya Pradesh, India

Karthika Rajendran
VIT School of Agricultural Innovations and Advanced Learning (VAIAL), Vellore Institute of Technology (VIT), Vellore, Tamil Nadu, India

M. Ammar Raza
Department of Field Crops, Faculty of Agriculture, Siirt University, Siirt, Turkey

Ayman El Sabagh
Department of Agronomy, Faculty of Agriculture, Kafrelsheikh University, Kafrelsheikh, Egypt; Department of Field Crops, Faculty of Agriculture, Siirt University, Siirt, Turkey

Diksha Saiyam
Metagenomics and Secretomics Research Laboratory, Department of Botany, Dr. Harisingh Gour Vishwavidyalaya (A Central University), Sagar, Madhya Pradesh, India

Abdul Salam
Zhejiang Key Lab of Crop Germplasm, Department of Agronomy, College of Agriculture and Biotechnology, Zhejiang University, Hangzhou, China

Aroona Saleem
Institute of Industrial Biotechnology, Government College University, Lahore, Pakistan

Rubab Shabbir
College of Agriculture, Fujian Agriculture and Forestry University, Fuzhou, China

Tayyab Shafiq
COMSATS University Islamabad (CUI), Islamabad, Pakistan

Satyawati Sharma
Indian Institute of Technology, CRDT, Hauz Khas, Delhi, India

Jay Shankar Singh
Department of Environmental Microbiology, School for Earth and Environmental Sciences, Babasaheb Bhimrao Ambedkar University, Lucknow, Uttar Pradesh, India

Meena Singh
Lab of Mushroom Biology, Department of Botany, Dr. Harisingh Gour University, Sagar, Madhya Pradesh, India

Sipan Soysal
Department of Field Crops, Faculty of Agriculture, Siirt University, Siirt, Turkey

Md. Tahjib-Ul-Arif
Plant Biology and Biofunctional Chemistry Lab, Department of Biochemistry and Molecular Biology, Bangladesh Agricultural University, Mymensingh, Bangladesh

Shobhit Raj Vimal
Ranjan Plant Physiology and Biochemistry Laboratory, Department of Botany, University of Allahabad, Prayagraj, Uttar Pradesh, India

Humaira Yasmin
COMSATS University Islamabad (CUI), Islamabad, Pakistan

Mohsin Zafar
COMSATS University Islamabad (CUI), Islamabad, Pakistan

Abbreviations

ABA	abscisic acid
ACC	1-aminocyclopropane-1-carboxylic acid
AHLs	N-acyl-homoserine lactones
AMF	arbuscular mycorrhizal fungi
ANOVA	analysis of variance
APX	ascorbate peroxidase
CA	catalase
cAPX	cytosolic APX
CE	capillary electrophoresis
CE-MS	capillary electrophoresis-based MS
CLPP	community-level physiological profiles
COSY	correlation spectroscopy
DAPG	2,4-diacetylphloroglucinol
DEPT	distortionless enhancement by polarization transfer
DGGE	denaturing gradient gel electrophoresis
DMDS	dimethyl sulfide
DSE	dark septate endophytic
EI	electron impact ionization
EMR	endophyte-mediated resistance
ePGPR	extracellular plant growth-promoting rhizobacteria
EPS	exopolysaccharides
ESI	electrospray ionization
FAO	Food and Agriculture Organization
GB	glycine betaine
GC	gas chromatography
GC-MS	gas chromatography-mass spectrometry
HCN	hydrogen cyanide
HMME	host-mediated microbiome engineering
HPLC	high-performance liquid chromatography
HRMS	high-resolution MS
HSQC	heteronuclear single quantum correlation spectroscopy
HTS	high-throughput screening techniques
IAA	indole acetic acid
iPGPR	intracellular plant growth-promoting rhizobacteria
ISR	induced systemic resistance
JA	jasmonic acid
LC	liquid chromatography

LC-MS	liquid chromatography-mass spectrometry
LPS	lipopolysaccharides
MAMP	microbe-associated molecular pattern
MAS-NMR	magic-angle spinning-NMR
MS	mass spectrometry
NC-Endophytes	non-clavicipitaceous endophytes
NLR	nucleotide-binding leucine-rich repeat receptors
NMDS	non-metric multidimensional scaling
NMR	nuclear magnetic resonance
NO	nitric oxide
NRPS	nonribosomal peptide synthetases
OP	osmoprotectants
PA	polyamines
PAL	phenylalanine-ammonia-lyase
PAMPs	pathogen-associated molecular patterns
PCA	phenazine-1-carboxylic acid
PCA	principal component analysis
PGP	plant growth-promoting
PGPB	plant growth-promoting bacteria
PGPM	plant growth-promoting microorganisms
PGPR	plant growth-promoting rhizobacteria
PHP	Pharmaceuticals and hygiene product
POD	peroxidase
PPO	polyphenol oxidase
PRRs	pattern recognition receptors
PSII	photosystem II
PTI	pattern-triggered immunity
Put	putrescine
QIIME	quantitative insights into microbial ecology
QS	quorum sensing
QS	quorum signaling
RAS/RT	root adhering soil to root tissue ratio
RFLP	restriction fragment length polymorphism
RNA-Seq	mRNA deep sequencing
ROS	reactive oxygen species
RZTs	root zone temperatures
SA	salicylic acid
SAR	systemic acquired resistance
SM	secondary metabolites
SOD	superoxide dismutase
SOLiD-SAGE	sequencing oligonucleotide ligate and detecting-serial analysis of gene expression
Spd	spermidine

Spm	spermine
SSH	suppression subtractive hybridization
ssNMR	solid-state NMR
TCA	tri-carboxylic acid
TOCSY	total correlation spectroscopy
TOF	time of flight
T-RFLP	terminal-restriction fragment length polymorphism
UPLC-MS	ultrahigh-pressure liquid chromatography-mass spectrometry
USEPA	United States Environmental Protection Agency
VOC	volatile organic compound

Preface

According to the United Nations Organization, global population is expected to top 10 billion in 2050, which will result in a growing demand for agricultural food and feed. Continuous increase in human population, deforestation, and the overuse of pesticides and chemical fertilizers have resulted in a reduction of cultivable land, crop growth, and soil health. Considering these issues without compromising the soil environment and human health is a major issue in the agricultural production sector and at the forefront of researchers. To achieve sustainable crop production without harming the environment, the role of plant-associated microbiome comes into existence.

Plants are naturally heavily loaded with beneficial microbiome that regulates essential functions in improving plant nutrition acquisition and provides protection against different biotic and abiotic stressors. The search for beneficial bacteria to aid in agricultural production, soil fertilization, disease management, and plant growth stimulation has become more important over the past few decades. Microbiomes interact with one another and contribute at many different interfaces, including those with plants, animals, and the soil environment, among others. The microbial population connected to the plants is referred to as the crop microbiome.

Previous research has been based on only culturable microbiome wherein researchers screened and isolated a particular bacterium and tested its plant growth promoting activities on a particular crop against individual stress. But due to the advancement in technologies like metagenomics, metabolomics, and various *in silico* tools, these phytomicrobiomes are easily applied/manipulated for the betterment of agricultural crop production.

Through this book, we tried to cover various aspects of plant microbiome research by inviting contributions from different scientists to provide the latest information available in this area of research.

In Chapter 1, Kumar introduces the concept of microbiome and its importance in crop plants, and discusses the techniques involved. He also covers the application of plant microbiome in sustaining agriculture productivity. In Chapter 2, Dubey et al. discuss general principles of microbial endophytes as warriors for plant defense against abiotic and biotic stress. They cover the role of endophytic bacteria and their multifarious interactions that are beneficial for plant growth and health. In Chapter 3, Dubey et al. highlight the importance of phyto-microbiomes: role in sustainable agriculture practices under abiotic stress. They discuss the approaches that necessitate a system-based approach to achieve the phytomicrobiome's potential in contributing to food security. In Chapter 4, Raj et al. present an overview of the role of metabolomics in bioremediation for sustainable crop production and the role of various metabolomics tools such as GC-MS, LC-MS, and NMR for the sustainable remediation of toxic pollutants from the environment. Moreover, this chapter discusses different web sources and software for metabolomics data interpretation.

In Chapter 5, Vimal et al. address the pressing issue of plant endomicrobiome services in global food security and agriculture sustainability. They also discuss the plant endomicrobiome diversity, mechanisms, services in cereal crop promotion, and stress toxicity management. In Chapter 6, Chowdhary et al. discuss the future perspectives of endophytic microbiome in sustainable agriculture.

In Chapter 7, Singh et al. thoroughly overview various abiotic and biotic stresses and their effects on crops, which will aid physiologists and molecular biologists in developing agronomically appropriate methods for the development of broad-spectrum stress-resistant crops. In Chapter 8, Afridi and Salam discuss the concept and strategies regarding the rhizosphere microbiome manipulation in crop improvement.

In Chapter 9, Javed et al. discuss the microbial traits in plant stress tolerance and their importance for agriculture. This chapter accentuates the beneficial traits of PGPR that should be kept in mind before their application in fields to gain maximum output as PGPR are highly specific to the soils and plant species. In Chapter 10, Sabagh et al. present an overview of the exogenous application of several compounds and phyto-microbiomes that mitigate the negative effects of salt stress on maize. In Chapter 11, Fatima et al. present the beneficial plant microbial association for sustainable agriculture. This chapter deals with the types of associations that develop between microorganisms and plants and how these interactions benefit each other. Moreover, the classification of PGP-rhizobacteria is also highlighted.

The current book offers details on current methods, technological advancements, the introduction of more recent technology, and areas that will be the focus of future research.

—*Editor*

Introduction

The global human population is predicted to cross 10 billion by 2050, and agricultural yield would need to grow by 60–100% to fulfill this increased demand for food. But to meet this demand on the limited lands and under different prevailing stress conditions is a big challenge. Furthermore, climate change and climate-related disasters directly impact biotic and abiotic factors that govern agroecosystems, compromising global food security. Resilient climate agriculture (CRA) focuses on the sustainable use of natural resources to boost agricultural productivity over the long term and in the face of climate variabilities.

Recently, the application of culturable and unculturable microbiome for solving these key challenges is gaining importance. Crop plants are heavily colonized by microorganisms collectively known as "microbiome" which establish a broad range of beneficial, neutral, and harmful interactions with their host. Associated microbiota, in the case of plants, plays a vital role in acquiring soil nutrients, tolerance to abiotic stresses, and disease control. These climate-resilient crops associated with these beneficial microbiotas are highly tolerant to abiotic and biotic stress and can withstand adverse climatic situations. Hence, there is an immediate demand at the regional, local, and international levels to develop sustainable agricultural practices by amalgamating conventional and modern microbiome-based technologies.

Consequently, it is critical to understand better the processes that determine microbial communities' composition and organization and the host's role in the recruitment and regulation of the microbiome that lives inside it. Modern techniques like genomics, metagenomics, proteomics, transcriptomics, metabolomics, fluxomics, etc., have given clearer vision about the functionality of these microbial communities at the various interfaces of the plants.

In this timely book, expert authors critically review and present their approaches and most important current research in this exciting field. The book provides information on existing methods, upgrading existing technologies, the advent of newer technologies, and focus areas for further development. This book covers subject areas related to agriculture, plant microbiome interaction, plant sciences, biotechnology, microbiome, metagenomics, soil science, environmental microbiology, genomics, proteomics, and metabolomics.

This book will be invaluable for training undergraduate and graduate (Masters and Doctoral) students, postdoctoral fellows, researchers, and libraries. It will also prove valuable for policymakers and the general audience interested in understanding the role of plant microbiomes in climate-resilient agriculture.

—Ashwani Kumar (PhD, IIT Delhi) FISEB, FNIE, YAI, FLS
Associate Professor,
Metagenomics and Secretomics Research Laboratory,
Department of Botany, University of Allahabad (A Central University),
Prayagraj, Uttar Pradesh, India

CHAPTER 1

Plant Microbiome Research: Tools and Techniques to Analyze Plant Microbiome for Sustainable Agriculture Productivity

ASHWANI KUMAR[1,2] and AMAN RAJ[2]

[1]*Metagenomics and Secretomics Research Laboratory, Department of Botany, University of Allahabad (A Central University), Prayagraj, Uttar Pradesh, India*

[2]*Metagenomics and Secretomics Research Laboratory, Department of Botany, Dr. Harisingh Gour University (A Central University), Sagar, Madhya Pradesh, India*

ABSTRACT

A microbiome is a community of microbes that inhabit a specific habitat. Like other living things, plants are inherently surrounded by various good and harmful microorganisms. The method of how microorganisms affect plant life under stressful circumstances has been kept secret for a very long time and is still not entirely understood. But as new omics tools (metagenomics, transcriptomics, metabolomics, and proteomics) and *in-silico* techniques were developed, microbiome research advanced and revolutionized our understanding while presenting renewed opportunities for sustainable agriculture. It has been demonstrated that the plant microbiome, a diverse community of microbes that live inside plants, is essential for plant development and health and protection from biotic and abiotic stresses. Therefore, it is crucial to develop sustainable agricultural practices that can maintain and improve crop output while lowering the impact of climate change and different pressures on the environment. Recent studies have demonstrated that the plant microbiome is essential for enhancing plant growth and resistance

Advances in Plant Microbiome Research for Climate Resilient Agriculture: Toward Sustainable Farming.
Ashwani Kumar (Ed.)

to various stressors, including disease, nutrient deprivation, and drought. In addition, plants and soil microbes can develop mutualistic partnerships that provide vital nutrients and defend them against deadly infections. The techniques and tools used to examine the plant microbiome, such as sample collection, DNA extraction, sequencing, and data analysis, are reviewed in this chapter. We also go over the difficulties and future directions in studying the plant microbiome, including the requirement for standardized procedures and extensive databases of microbial genomes and competencies. We can learn more about the plant microbiome and its potential to increase agriculture sustainability as microbiome analysis advances.

1.1 INTRODUCTION

The rapidly expanding world population, rapid industrialization and urbanization, and anthropogenic activities have affected the terrestrial system and have put a lot of burden on the ecosystem and the hidden players regulating it (Singh et al., 2017). As per the reports of the United Nations, the global human population is expected to exceed approx. 10 billion by the year 2050, thus putting more pressure on land to feed such huge masses (Muneer, Dubey, Raj et al., 2022). Moreover, the increased world population will lead to several human interventions in ecosystem functioning due to deforestation, thus affecting the soil water holding capacity and enhancing its erosion. Furthermore, the world needs another green revolution and this will come at the cost of extensive uses of chemical fertilizers and pesticides, thus affecting the soil health and the hidden goldmines of the soil, i.e., the soil and plant microbiomes (Raj & Kumar, 2022b). Multiple strategies are required to address these issues, including using microbes that promote plant growth as biostimulants to raise crop productivity. The use of microbial consortia or single microorganisms as inoculum was previously explored, and biostimulants' application in agriculture is not new (Narayanasamy et al., 2015). Microbial inoculants can be successfully transferred from the lab to the field, but this is still difficult. Numerous species of crops and a variety of crops, varying environmental conditions between fields, and the exponential rise in the number of microbial isolates are the leading causes of this. Therefore, a comprehensive strategy for the use of "biostimulants" is required, including "diagnostics" of the desired crop's genotype and the field environment (e.g., soil), the best agricultural practices, screening for inoculations from available culture collections, expanding scientific research on microbiomes, and finally incorporating all of those mentioned above into large-scale industrial production (Arora & Mishra, 2016). Numerous researchers have studied microbial inoculants since the 1980s but have had only patchy success. With further knowledge of plant microbiota concerning biotic and abiotic stressors, plant genotype, and environmental variables, it may be possible to identify candidates or methods for inoculation that are more suited for a given environment (Naz et al., 2022). Fungi, archaea, and bacteria constitute only a few of the diverse types of organisms that make up the plant microbiome (Jacoby et al., 2017; Saad et al., 2020). The vast community of microorganisms that dwell in and on plants is called the microbiome of plants. These microorganisms are critical for maintaining agricultural output since they are crucial for plant growth and health (Raj et al., 2023). To better understand the interactions between plants and the accompanying microbes, there has been an increase in interest in investigating plant microbiomes in recent years. To sustain agricultural output, this chapter will give an overview of the techniques and tools currently employed to study plant microbiomes.

1.2 PLANT MICROBIOME

A relatively recent area of research called "plant microbiome" focuses on the many communities of microbial organisms that live inside and outside of plant tissues and support their growth and health. These include microbes that interact intricately with plants, such as bacteria, fungi, viruses, and others (Chouhan et al., 2021). A plant microbiome is a holobiome comprising a phyllosphere microbiome, endosphere microbiome, spermosphere microbiome, rhizosphere microbiome, and the soil-associated non-rhizospheric microbiome (Jat et al., 2021). Recent studies have demonstrated that the plant microbiome is essential for increasing plant growth and resistance to various stressors, including disease, nutrient deprivation, and drought. Plants and soil microbes can develop mutualistic partnerships that provide vital nutrients and defend them against deadly infections (Dubey, Malla et al., 2020; Gontia-Mishra et al., 2016). It is well recognized that microbial populations are associated with plants and soil that positively affect plant health, soil stability, and sustainability in both agriculture and natural ecosystems. Beneficial soil bacteria assist plants in the uptake of additional nutrients from the soil and provide resistance to biotic and abiotic stresses (Saleem et al., 2017).

In return, microorganisms use the waste byproducts from plants as a source of energy. The microbiome also aids in the defense against pathogens through niche competition, the production of antibiotics, ensuring the accessibility of nutrients, the release of secondary metabolites (SMs), and triggering the development of systemic resistance that activates the plant defense system against initial pathogen attack (Santos & Olivares, 2021). The microbiome also helps the plants by producing enzyme activity, which can lessen pathogens' harmful effects by making antimicrobials like hydrogen cyanide (HCN) and siderophore (Dubey, Kumar et al., 2020; Jacoby et al., 2017; Kumar & Dubey, 2020). However, the relationship between beneficial microbiota and plants has been impacted due to human interference, such as selective breeding, ongoing monoculture, and dumping of agrochemicals into agroecosystems. With a long history of domestication and strict breeding, cotton and maize have high input requirements and are rife with pests and illnesses that are thought to be caused by the loss of beneficial microbiota. The efficient use of selected microbes or microbial consortia towards yield maximization may result from a better understanding of the function of microorganisms on plant growth and stress response (White et al., 2019). In general, the study of the plant microbiome is a fascinating, promptly developing subject matter that has the potential to completely change how we see plant biology and agriculture. Figure 1.1 shows the overview of the plant microbiome.

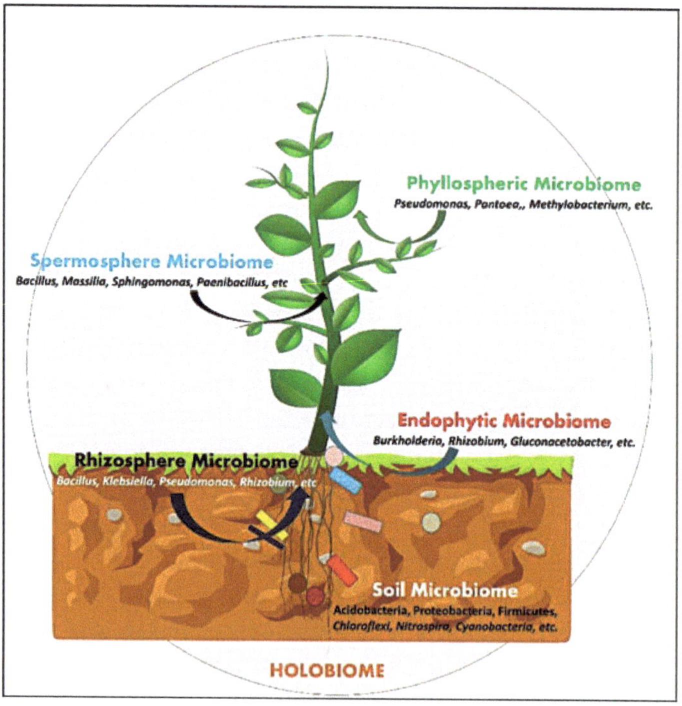

FIGURE 1.1 An overview of plant microbiome.

1.3 PLANT MICROBIOME AND ITS ROLE IN AGRICULTURAL SUSTAINABILITY

Because of the continued strain that climate change and population increase are putting on our food systems, agricultural sustainability is a subject that is receiving more and more attention (Ji et al., 2023). The plant microbiome, which refers to the diverse community of microbes that live on and inside plants, is one area of interest in enhancing agricultural sustainability (Dhankher & Foyer, 2018). The plant microbiome plays a crucial role in nutrient cycling and soil health. Through mineralization and decomposition processes, microbes in the plant microbiome can aid in nutrient cycling, increase nutrient availability, and facilitate nutrient uptake from the soil. This can lessen the need for synthetic fertilizers and enhance soil health, improving crop yields and less negatively impacting the environment. Additionally, for long-term agricultural sustainability, the plant microbiome can aid in preserving soil structure and reducing erosion (Yergeau et al., 2015). Additionally, the microbes in the plant microbiome can aid in plants' growth and disease resistance. While some microbes can cause systemic resistance in plants, making them more resistant to pathogens and pests, other microbes can produce hormones that promote plant growth. This can enhance crops' general well-being and productivity while reducing the need for chemical pesticides (Ruzzi et al., 2016; Walvekar et al., 2017). Microbial inoculants, which can be added to agricultural soils to enhance plant growth and nutrient uptake, are preparations of beneficial microbial organisms. Microbial inoculants have several benefits over conventional inputs and can be utilized as a natural substitute for synthetic fertilizers and chemical pesticides. For instance, microbial inoculants can increase soil health and fertility and are economical and environmentally benign. The effectiveness of microbial inoculants in encouraging plant development and disease resistance can also be improved because they can be specifically adapted to different crop types and growing environments (Saad et al., 2020; Santos & Olivares, 2021). Even though using the plant microbiome to increase agricultural sustainability may be advantageous, several issues still need to be resolved. One obstacle is the plant microbiome's complexity, making it challenging to pinpoint the most beneficial microorganisms for crops and growth environments. Additionally, the soil characteristics and crop management techniques, which can differ greatly across various agricultural systems, can significantly impact the efficacy of microbial inoculants. Finally, legalizing microbial inoculants can be difficult due to the absence of industry regulation and standardization.

1.3.1 PLANT MICROBIOME AS FERTILIZERS

Modern agriculture's usage of synthetic fertilizers has had many detrimental effects on the environment, including soil erosion, waterway pollution, and greenhouse gas emissions. The use of plant microbiomes as a natural substitute for synthetic fertilizers has drawn more attention in recent years. A wide variety of microorganisms that can support plant growth and nutrient uptake make up the microbiome of plants (Kumar & Dubey, 2020). Interactions between plants and microbes are essential for plants to absorb nutrients. The accessibility of nutrients for plant growth can be improved by the microbes in the plant microbiome, which can help nutrient uptake from the soil. The following are some of the ways that microorganisms can enhance nutrient uptake:

1. **Nitrogen Fixation:** Some microorganisms in the plant microbiome, such as nitrogen-fixing bacteria (*Rhizobium*, *Azotobacter*, *Azospirillum*, *Bradyrhizobium*, etc.), can transform atmospheric nitrogen into a form that plants can use. This method can make more nitrogen available for plant growth, requiring less synthetic nitrogen fertilizer (Tkacz & Poole, 2021).
2. **Phosphorus Solubilization:** Certain microorganisms in the plant microbiome, including phosphorus-solubilizing bacteria and fungi, can liberate phosphorus from soil particles, making it accessible for plant absorption (Chakraborty et al., 2013). For ex: *Bacillus*, *Pseudomonas*, *Kleibsiella*, *Aspergillus*, *Micrococcus*, *Fusarium*, *Rhizobium*, etc.
3. **Siderophore Production:** Siderophores are tiny compounds capable of chelating iron and making it available for plant uptake. A few microorganisms, like *Pseudomonas fluorescence* in the plant microbiome (Braga et al., 2016).

Employing several methods, including hormone production, induced systemic resistance (ISR), and biofilm formation, microbes in the plant microbiome can also stimulate plant growth. Auxins (*Paenibacillus*, *Bacillus*, *Pseudomonas*, *Bradyrhizobium*, etc.), cytokinins (*Azospirillum*) (Zaheer et al., 2022), and gibberellins (*Rhizobium meliloti*, *Azospirillum* sp., *Gibberella fujikuroi*, *Bradyrhizobium japonicum*, etc.) (Bottini et al., 2004), which are plant growth hormones produced by some microorganisms in the plant microbiome, can help plants grow and develop, while other microbes in the plant microbiome can make plants develop systemic resistance (*Bacillus*, *Pseudomonas*, *Trichoderma*, and mycorrhiza species) (Pieterse et al., 2014), which increases their tolerance to infections and pests. Several additional microbes may develop biofilms on plant roots as part of the plant microbiome, such as *Paenibacillus polymyxa,* which can improve nutrient uptake, shield plants from environmental stress, and encourage plant growth (Timmusk et al., 2005). Microbial inoculants, which can be added to agricultural soils to enhance plant growth and nutrient uptake, are preparations of beneficial bacteria and fungi. Microbial inoculants are a natural substitute for synthetic fertilizers and offer several benefits over conventional fertilizers, including economic effectiveness, sustainability, and environmental friendliness (Sharma et al., 2021).

Regardless of the potential advantages of using plant microbiomes as fertilizers, several issues need to be resolved. For example, given that plant microbiomes are so diverse, it can be challenging to pinpoint the most helpful microbes for plant growth. In addition, because the soil's conditions can significantly influence microbial inoculants' effectiveness, their application may need to be customized for particular cropping systems and soil types. In addition, there is currently a lack of industry regulation and standardization, which makes it a daunting task to commercialize microbial inoculants.

1.3.2 PLANT MICROBIOME IN REGULATING ABIOTIC STRESSES

Abiotic stress, including drought, salinity, and extremely high or low temperatures, is a significant problem for crop production worldwide. The numerous microbial communities in plant tissues and the rhizosphere comprise the plant microbiome, which has become a key regulator of plant responses when subjected to abiotic stress. Drought stress is among the most significant abiotic factors affecting crop yield worldwide. By controlling stomatal conductance, promoting osmoprotectant synthesis, and enhancing root growth, microbes in the plant microbiome can increase the efficiency with which plants use water. For instance, certain *Rhizobacteria* and fungal endophytes can improve plants' ability to withstand drought by generating phytohormones and inducing the expression of genes relevant to stress (Miranda et al., 2023; Poudel et al., 2021). Another significant abiotic stress that restricts crop productivity is salinity stress, brought on by elevated salt levels in the soil. By decreasing the uptake of noxious ions and improving food availability, microbes in the plant microbiome can help plants endure salinity stress. For instance, certain rhizobacteria (*Pseudomonas* spp., *Serendipita indica, Bacillus* spp., and arbuscular mycorrhizal fungi (AMF)) can boost the availability of nutrients like nitrogen and phosphorus, which are crucial for plant growth, while decreasing the accumulation of sodium ions in plants (Backer et al., 2018). Extreme temperature stress can also impact crop yield, encompassing both hot and low temperatures. By controlling plant physiology and metabolism, microbes in the plant microbiome can assist plants in withstanding the stress caused by extreme temperatures. By boosting antioxidant activity and shielding plant tissues from heat damage, for instance, some fungal endophytes and *Enterobacter sp. SA187* can help plants withstand high temperatures (Dastogeer et al., 2022). Plants' ability to withstand abiotic stress can be improved by applying microbial inoculants. Microbial inoculants can improve plant growth and stress tolerance by enhancing plant-microbe interactions, regulating plant hormone levels, and improving nutrient uptake. For instance, inoculating crops like wheat, maize, and soybean with certain rhizobacteria can enhance plant growth and drought tolerance (Bittencourt et al., 2023). Overall, it can be concluded that the plant microbiome is crucial in controlling how plants react to abiotic stress, such as drought, salinity, and high temperatures. The sustainability of agriculture and the harmful environmental effects of existing agricultural practices may be improved by developments in plant microbiome research and practices.

1.3.3 PLANT MICROBIOME IN REGULATING BIOTIC STRESSES

Pest and pathogen attacks generate biotic stress, a significant global concern for crop productivity. The numerous microbial communities that live in plant tissues and the rhizosphere comprise the plant microbiome, which has become a key regulator of plant responses to biotic stress. Pest insects are a severe biotic stressor that can drastically lower crop output. Plant microbes can enhance insect pest control by generating insecticidal substances (for ex. *Bacillus thuringiensis* produces Bt-toxin) or by improving plant defenses. For instance, some bacterial endophytes and fungi may create substances that are poisonous to insects, while some bacteria help prepare plants' insect defenses (Chattopadhyay et al., 2017; Munir et al., 2022). Fungus infections like *Fusarium* and *Phytophthora* species can bring on significant crop losses. By developing antifungal chemicals, promoting plant resistance, or competing for nutrients and space, microbes in the plant microbiome can aid in managing fungal diseases. For instance, certain bacteria (*Bacillus, Pseudomonas, Klebsiella*, etc.) can compete with certain fungi for resources, and certain bacterial and fungal endophytes may produce antifungal substances that prevent the growth of fungi that cause disease (Khan et al., 2018). Agriculture can benefit from biocontrol agents, microorganisms that can control infections and pests. By enhancing nutrient uptake, controlling plant hormone levels, and improving plant-microbe interactions, biocontrol agents can improve plant growth and control biotic stress. Certain fungi and bacteria can be utilized as biocontrol agents to control bacterial and fungal infections in crops like tomato, cucumber, and pepper.

1.4 BIOINFORMATICS APPROACH FOR PLANT MICROBIOME ANALYSIS

The complex and varied community of microbes known as the plant microbiome interacts with and is crucial to its host plant's growth, development, and stress resistance. Recent breakthroughs in sequencing technology and bioinformatics have made it possible for researchers to investigate the structure and function of the microbiome more precisely and extensively. This has accelerated the study of plant microbiomes (Berendsen et al., 2012; Lakshmanan et al., 2014; Rascovan et al., 2016). The research of plant microbiomes depends on sequencing technologies since they make it possible to sequence microbial DNA and RNA at high throughput from complicated specimens from the environment. Metagenomic sequencing, Amplicon sequencing, and meta-transcriptomic sequencing are some available sequencing techniques (Gutleben et al., 2017; Hemmat-Jou et al., 2018; Muneer et al., 2018). Through the use of PCR, specific microbial DNA regions are amplified, and the amplicons produced are then sequenced (Enespa & Chandra, 2022). This method can deliver high-resolution data on the diversity of microbes and the composition of communities while being relatively affordable. Sequencing a sample's entire DNA composition, comprising microbial and host DNA, is known as metagenomic sequencing (Handelsman, 2004). This method can reveal information on the functional abilities of microbes and the existence of particular genes involved in interactions between plants and microbes. Sequencing of all RNA, including microbial and host RNA, found in a sample is known as meta-transcriptomic sequencing. This strategy can reveal details about the functional activity of the microbiome and microbial gene expression (Vanessa Aguiar-Pulido et al., 2016). To properly analyze plant microbiomes, microbiome sequencing data must be processed. Low-quality reads adaptor sequences and host DNA must all be removed from the raw sequencing data through filtering. Quality control is crucial to guarantee the accuracy and dependability of sequencing data.

After quality control, reads are categorized according to taxonomic classifications, and the relative abundance of each taxon is determined.

Numerous bioinformatics tools, including QIIME, mothur, WebMGA, and MG-RAST, are extensively used to accomplish this (Kuczynski et al., 2012; Schloss et al., 2009; Silva et al., 2013; Wu et al., 2011). Then, statistical analysis is performed to distinguish microbiome composition and function changes between samples and find relationships between microbiome traits and host phenotypes like plant growth or stress tolerance. Various statistical techniques, such as ordination, hypothesis testing, and artificial intelligence, are available to investigate the plant microbiome. To see variations in microbiome composition between samples, ordination techniques like principal component analysis (PCA) and non-metric multidimensional scaling (NMDS) are used (Wolinska et al., 2020). The analysis of variance (ANOVA) and Kruskal-Wallis tests, among others, can be employed to detect significant variations in the microbiome composition of sample groups (Mansouriieh et al., 2019). Finally, plant phenotypes are predicted using machine learning techniques like random forests and support vector machines, which consider the microbiome's structure and function (Misra et al., 2019).

1.4.1 FROM SAMPLE COLLECTION TO SEQUENCING

Advances in bioinformatics and sequencing technologies have revolutionized the study of microbiomes, allowing scientists to learn more about the structure and function of microbial communities (Raj & Kumar, 2022a; Sharma et al., 2022). The quality and quantity of DNA extracted from the samples might impact subsequent analyses; hence collecting samples is an important stage in microbiome analysis (Handelsman, 2004). Depending on the environment being investigated, sample collection techniques will vary, but in general, they entail taking samples while employing sterile procedures to reduce contamination from outside sources. For instance, samples can be obtained from numerous body regions using swabs, fecal collection devices, or tissue biopsies in human microbiome studies. In soil and plant microbiome studies, samples can be taken utilizing soil cores, rhizosphere sampling, or plant tissue homogenization (Saad et al., 2020). DNA extraction must be done after the sample has been collected to separate the microbial DNA from the host and ambient DNA. Depending on the sample type, many DNA extraction techniques can be used, although mechanical and chemical lysis is often used to liberate DNA from cells. Various commercial DNA extraction kits are available for microbiome analysis, including the MoBio PowerMag Microbial DNA Isolation Kit for human microbiome samples and the Qiagen DNeasy PowerSoil Kit for soil and plant samples. Due to the potential effects of low DNA yield or contamination on subsequent analysis, it is crucial to ensure that DNA extraction techniques are optimized to maximize the yield and purity of microbial DNA (Phulpoto et al., 2020). Sequencing is done after DNA extraction to produce significant data on the diversity of microbes and community structure. Amplicon sequencing, shotgun metagenomic sequencing, and meta-transcriptomic sequencing are a few of the sequencing techniques used for microbiome investigation (Muneer, Dubey, Kumar et al., 2022; Mendes & Tsai, 2018). Amplicon sequencing utilizes PCR to amplify certain microbial DNA sections and sequence the resulting amplicons. This method can deliver high-resolution data on microbial diversity and community composition while relatively affordable. Shotgun metagenomic sequencing entails sifting through every piece of DNA in a sample, including microbial and host DNA. This method can reveal information about the potential for certain genes associated with microbial function and their presence. Sequencing of all RNA, including microbial and host RNA, found in a sample is known as meta-transcriptomic sequencing (Aguiar-Pulido et al., 2016). Bioinformatics analysis is then carried out to handle the massive volumes of data generated after sequencing. This requires several procedures, including statistical analysis, taxonomy classification, and quality control.

1.4.2 TOOLS FOR MICROBIOME ANALYSIS

For the differentiation and identification of microorganisms, a variety of methods/procedures are available, including isolation, the purification process of visible colonies, and microbe rearing. However, all of these methods are used in the laboratories, and they are outdated. Under a microscope, many bacteria resemble one another and won't survive outside their natural environments (Ferone et al., 2020). Only about 1% of microbes have been uncovered due to a lack of methods and instruments. This is because some bacteria are viable but non-culturable, whilst others are found in settings with high pH, pressure, temperature, salinity, and other conditions (Raj et al., 2021). Concerning this, early scientists only looked into a few types of microorganisms and a specific area of microbial habitat includes microbes that have been cultivated in a lab. However, the advancements in molecular biology and the emergence of a culture-independent approach made it possible to know the unknown and explore the unexplored (Bodor et al., 2020). Aquatic populations can be identified without any data by using some molecular techniques, such as G + C%, RFLP (restriction fragment length polymorphism), DNA hybridization, and community-level physiological profiles (CLPP) (Ritchie et al., 2000). The determination of microbial species by retaining polymerase chain reaction from samples taken from the environment has resulted in the use of a variety of DNA fingerprinting techniques, including terminal-restriction fragment length polymorphism (T-RFLP), temperature gradient gel electrophoresis (TGGE), repetitive PCR, and denaturing gradient gel electrophoresis (DGGE) (Portillo et al., 2011).

DNA microarray analysis is a superb method for understanding the soil microbiome. This method is employed to study the genetic makeup of various regions of an organism's genome or to simultaneously assess the degree of gene expression at multiple loci. In this method, DNA from the sample is extracted and amplified using PCR. Additional tagged DNA samples are examined using a microarray. The microarray has a wide variety of oligonucleotide probes, ranging from 10 to 1,000 (Enespa & Chandra, 2022). The 16S rRNA or a functional gene fragment could be used as a probe. For the probes on the chip to bind to or hybridize with the DNA samples, they need to be identical. After binding, the chip's signals are digitally examined (Bumgarner, 2013). The emergence of metagenome provides a clear picture of genetic diversity, species composition, species evolution, and interspecies interactions. Before the last few decades, several complex omics sequencing methods were developed, including Illumina sequencing and pyrosequencing. Research on metagenomics and meta-transcriptomics can benefit from these techniques (Handelsman, 2004; Rodríguez et al., 2015; Yoo et al., 2017). The proficiency and objectivity of next-generation sequencing, when obtained directly from a soil sample, are examined. Therefore, the collection of DNA from the soil sample determines how the results are represented and how biased the conclusions are. The Illumina sequencer may be used to fill in any gaps in the 454 produced sequence data due to its high throughput, even if the 454 sequencers are directly employed for the generation of base readings because it fosters longer reads (Bhattacharyya et al., 2018; Muccee & Ejaz, 2020; Wang et al., 2014).

Bioinformatics-based plant microbiome analysis is a rapidly developing field with many issues that must be resolved. Standardized microbiome sequencing and data processing techniques must be developed to improve reproducibility and comparability between investigations (Raj et al., 2023). A larger database of microbial genomes and functions is required to improve the taxonomic and functional annotation of microbiome sequencing data. Finally, further study is required to develop machine learning methods that can precisely predict plant phenotypes based on microbiome data. Table 1.1 shows the various tools used for microbiome analysis.

TABLE 1.1 Bioinformatics Tools/Pipelines Used for Microbiome Analysis

SL. No.	Bioinformatic Tool/Software	Description	Type of Platform	Link/Website	References
1.	Microbiome analyst	A comprehensive statistical, graphical, and meta-analytical tool for examining microbiome data.	Web-based	http://www.microbiomeanalyst.ca/	Dhariwal et al. (2017)
2.	EBI	To compare functional analyses of sequence.	Web-based	https://www.ebi.ac.uk/metagenomics/	Madeira et al. (2022)
3.	GraPhlAn	Produces microbial genomes and metagenomes in high-quality, compact visualizations.	Web-based	http://segatalab.cibio.unitn.it/tools/graphlan	Asnicar et al. (2015)
4.	deFUME	It is a simple-to-use web service for trimming, assembly, and functional annotation of Sanger sequencing data produced from functional selection experiments.	Web-based	http://www.cbs.dtu.dk/services/deFUME/	Van Der Helm et al. (2015)
5.	METAVIR	Annotate viral metagenomics sequences (raw reads assembled contigs).	Web-based	http://metavir-meb.univbpclermont.fr/	Roux et al. (2011)
6.	Phinch	It offers a user-friendly visualization tool that enables interactive exploration and manipulation of big biological datasets.	Web-based	http://phinch.org/	Bik & Inc. (2014)
7.	PICRUSt	A computational method for 16S amplicon sequencing-based functional metagenome profiling.	Stand-alone	http://picrust.github.com/	Douglas et al. (2020)
8.	METAGENassist	An accessible web server for comparative metagenomic analysis is called METAGENassist. Large-scale comparisons of the genomic or taxonomic census data from bacterial samples collected from various settings are a key component of comparative metagenomic studies.	Web-based	http://www.metagenassist.ca/METAGENassist	Arndt et al. (2012)
9.	WebMGA	A customizable web server for metagenomic analysis.	Web-based	http://weizhongli-lab.org/metagenomic-analysis/	Wu et al. (2011)
10.	MEGAN	Diversity analysis and visualization (needs similarity alignments as input).	Stand-alone	http://ab.inf.uni-tuebingen.de/software/megan	Yu & Zhang (2012)

TABLE 1.1 *(Continued)*

SL. No.	Bioinformatic Tool/Software	Description	Type of Platform	Link/Website	References
11.	QIIME	Visualization, diversity analysis, and data truncation and filtering.	Stand-alone	http://qiime.org/	Kuczynski et al. (2012)
12.	SmashCommunity	Performs assembly and gene prediction mainly for data from Sanger and 454 sequencing technologies.	Stand-alone	http://www.bork.embl.de/software/smash/	Arumugam et al. (2010)
13.	Mothur	Fast processing of large sequence data.	Stand-alone	http://www.mothur.org/	López-García et al. (2018)
14.	MG-RAST	A high-throughput pipeline for functional metagenomic analysis.	Webserver	http://metagenomics.anl.gov/	Oliveira et al. (2021)
15.	VAMPS	An integrated tool used for data visualization and analysis obtained for the microbial population.	Webserver	https://vamps2.mbl.edu	Huse et al. (2014)

1.5 CONCLUSION

A relatively recent area of research called "plant microbiome" focuses on the many communities of microorganisms that live in plant tissues and support their growth and health. This includes microbes that interact intricately with plants, such as bacteria, fungi, viruses, and others. The plant microbiome contributes to the soil's health, plant development, and disease resistance, all of which are crucial for agricultural sustainability. The sustainability of agriculture and the harmful environmental effects of existing agrarian practices may be improved by developments in plant microbiome research and practices. Plant microbiome analysis using bioinformatics approaches is a rapidly evolving field with several challenges. One is the need for more comprehensive databases of microbial genomes and functions to improve taxonomic and functional annotation of microbiome sequences. The development of more precise and dependable methods for taxonomic and functional annotation of sequencing data, the integration of multi-omics data to enhance our comprehension of microbiome function, and the implementation of machine learning techniques that can precisely predict plant phenotypes based on microbiome data will be the focus of future bioinformatics approaches for plant microbiome analysis. Our understanding of interactions between the plant and the microbiome will advance due to advances in bioinformatics techniques.

KEYWORDS

- **abiotic stresses**
- **agricultural sustainability**
- **biotic stresses**
- **drought**
- **fertilizers**
- **microbiome**
- **omics**
- **plant growth**
- **sustainable agriculture**

REFERENCES

Aguiar-Pulido, V., Huang, W., Suarez-Ulloa, V., Cickovski, T., Mathee, K., & Narasimhan, G. (2016). Metagenomics, metatranscriptomics, and metabolomics approaches for microbiome analysis: supplementary issue: bioinformatics methods and applications for big metagenomics data. *Evolutionary Bioinformatics, 12,* EBO-S36436. https://doi.org/10.4137/EBO.S36436.

Aguiar-Pulido, V., Huang, W., Suarez-Ulloa, V., Cickovski, T., Mathee, K., & Narasimhan, G., (2016). Metagenomics, metatranscriptomics, and metabolomics approaches for microbiome analysis. *Evolutionary Bioinformatics*, *12*, 5–16. https://doi.org/10.4137/EBO.S36436.

Arndt, D., Xia, J., Liu, Y., Zhou, Y., Guo, A. C., Cruz, J. A., Sinelnikov, I., et al., (2012). METAGENassist: A comprehensive web server for comparative metagenomics. *Nucleic Acids Research*, *40*(W1). https://doi.org/10.1093/nar/gks497.

Arora, N. K., & Mishra, J., (2016). Prospecting the roles of metabolites and additives in future bioformulations for sustainable agriculture. In: *Applied Soil Ecology*. https://doi.org/10.1016/j.apsoil.2016.05.020.

Arumugam, M., Harrington, E. D., Foerstner, K. U., Raes, J., & Bork, P., (2010). Smash community: A metagenomic annotation and analysis tool. *Bioinformatics*, *26*(23), 2977, 2978. https://doi.org/10.1093/BIOINFORMATICS/BTQ536.

Asnicar, F., Weingart, G., Tickle, T. L., Huttenhower, C., & Segata, N., (2015). Compact graphical representation of phylogenetic data and metadata with GraPhlAn. *PeerJ*, *3*, e1029. https://doi.org/10.7717/peerj.1029.

Backer, R., Rokem, J. S., Ilangumaran, G., Lamont, J., Praslickova, D., Ricci, E., Subramanian, S., & Smith, D. L., (2018). Plant growth-promoting rhizobacteria: Context, mechanisms of action, and roadmap to commercialization of biostimulants for sustainable agriculture. *Frontiers in Plant Science*, *871*, 1473. https://doi.org/10.3389/FPLS.2018.01473/BIBTEX.

Berendsen, R. L., Pieterse, C. M. J., & Bakker, P. A. H. M., (2012). The rhizosphere microbiome and plant health. In: *Trends in Plant Science*. https://doi.org/10.1016/j.tplants.2012.04.001.

Bhattacharyya, D., Duta, S., Yu, S. M., Jeong, S. C., & Lee, Y. H., (2018). Taxonomic and functional changes of bacterial communities in the rhizosphere of Kimchi cabbage after seed bacterization with *Proteus vulgaris* JBLS202. *Plant Pathology Journal*, *34*(4), 286–296. https://doi.org/10.5423/PPJ.OA.03.2018.0047.

Bik, H. M., & Pitch Interactive Inc. (2014). Phinch: an interactive, exploratory data visualization framework for–Omic datasets. BioRxiv, 009944. https://doi.org/10.1101/009944.

Bittencourt, P. P., Alves, A. F., Ferreira, M. B., Da Silva, I. L. E. S., Pinto, V. B., & Olivares, F. L., (2023). Mechanisms and applications of bacterial inoculants in plant drought stress tolerance. *Microorganisms, 11*(2), 502. https://doi.org/10.3390/MICROORGANISMS11020502.

Bodor, A., Bounedjoum, N., Vincze, G. E., Erdeiné, K. Á., Laczi, K., Bende, G., Szilágyi, Á., et al., (2020). Challenges of unculturable bacteria: Environmental perspectives. *Reviews in Environmental Science and Bio/Technology*, *19*(1), 1–22. https://doi.org/10.1007/S11157-020-09522-4.

Bottini, R., Cassán, F., & Piccoli, P., (2004). Gibberellin production by bacteria and its involvement in plant growth promotion and yield increase. *Applied Microbiology and Biotechnology*, *65*(5), 497–503. https://doi.org/10.1007/S00253-004-1696-1.

Braga, R. M., Dourado, M. N., & Araújo, W. L., (2016). Microbial interactions: Ecology in a molecular perspective. *Brazilian Journal of Microbiology*, *47*, 86–98. https://doi.org/10.1016/j.bjm.2016.10.005.

Bumgarner, R., (2013). Overview of DNA microarrays: Types, applications, and their future. *Current Protocols in Molecular Biology, 101*(1), 22.1.1–22.1.11. https://doi.org/10.1002/0471142727.MB2201S101.

Chakraborty, U., Chakraborty, B. N., Chakraborty, A. P., Sunar, K., & Dey, P. L., (2013). Plant growth promoting rhizobacteria mediated improvement of health status of tea plants. *Indian Journal of Biotechnology*.

Chattopadhyay, P., Banerjee, G., & Mukherjee, S., (2017). Recent trends of modern bacterial insecticides for pest control practice in integrated crop management system. *3 Biotech, 7*(1), 1–11. https://doi.org/10.1007/S13205-017-0717-6/FIGURES/1.

Chouhan, G. K., Verma, J. P., Jaiswal, D. K., Mukherjee, A., Singh, S., De Araujo, P. A. P., Liu, H., et al., (2021). Phytomicrobiome for promoting sustainable agriculture and food security: Opportunities, challenges, and solutions. *Microbiological Research, 248*, 126763. https://doi.org/10.1016/j.micres.2021.126763.

Dastogeer, K. M. G., Zahan, M. I., Rhaman, M. S., Sarker, M. S. A., & Chakraborty, A., (2022). Microbe-mediated thermotolerance in plants and pertinent mechanisms- a meta-analysis and review. *Frontiers in Microbiology, 13*. https://doi.org/10.3389/FMICB.2022.833566/FULL.

Dhankher, O. P., & Foyer, C. H., (2018). Climate resilient crops for improving global food security and safety. In: *Plant Cell and Environment*. https://doi.org/10.1111/pce.13207.

Dhariwal, A., Chong, J., Habib, S., King, I. L., Agellon, L. B., & Xia, J., (2017). Microbiome analyst: A web-based tool for comprehensive statistical, visual, and meta-analysis of microbiome data. *Nucleic Acids Research, 45*(W1), W180–W188. https://doi.org/10.1093/nar/gkx295.

Douglas, G. M., Maffei, V. J., Zaneveld, J. R., Yurgel, S. N., Brown, J. R., Taylor, C. M., Huttenhower, C., & Langille, M. G. I., (2020). PICRUSt2 for prediction of metagenome functions. *Nature Biotechnology, 38*(6), 685–688. https://doi.org/10.1038/s41587-020-0548-6.

Dubey, A., Kumar, A., & Khan, M. L., (2020). Role of biostimulants for enhancing abiotic stress tolerance in fabaceae plants. In: *The Plant Family Fabaceae* (pp. 223–236).

Dubey, A., Malla, M. A., Kumar, A., Dayanandan, S., & Khan, M. L., (2020). Plants endophytes: Unveiling hidden agenda for bioprospecting toward sustainable agriculture. In: *Critical Reviews in Biotechnology* (Vol. 40, No. 8, pp. 1210–1231). Taylor & Francis. https://doi.org/10.1080/07388551.2020.1808584.

Egamberdieva, D., Shrivastava, S., & Varma, A. (Eds.). (2015). Plant-growth-promoting rhizobacteria (PGPR) and medicinal plants (Vol. 42). Cham: Springer International Publishing. https://link.springer.com/book/10.1007/978-3-319-13401-7.

Enespa, & Chandra, P. (2022). Tool and techniques study to plant microbiome current understanding and future needs: an overview. *Communicative & Integrative Biology, 15*(1), 209-225. https://doi.org/10.1080/19420889.2022.2082736.

Ferone, M., Gowen, A., Fanning, S., & Scannell, A. G. M., (2020). Microbial detection and identification methods: Bench top assays to omics approaches. *Comprehensive Reviews in Food Science and Food Safety, 19*(6), 3106–3129. https://doi.org/10.1111/1541-4337.12618.

Gontia-Mishra, I., Sapre, S., Sharma, A., & Tiwari, S., (2016). Amelioration of drought tolerance in wheat by the interaction of plant growth-promoting rhizobacteria. *Plant Biology, 18*(6), 992–1000. https://doi.org/10.1111/plb.12505.

Gutleben, J., Chaib De Mares, M., Van Elsas, J. D., Smidt, H., Overmann, J., & Sipkema, D. (2018). The multi-omics promise in context: from sequence to microbial isolate. Critical reviews in microbiology, 44(2), 212-229. https://doi.org/10.1080/1040841X.2017.1332003.

Handelsman, J., (2004). Metagenomics: Application of genomics to uncultured microorganisms. *Microbiology and Molecular Biology Reviews*, *68*(4), 669–685. https://doi.org/10.1128/MMBR.68.4.669-685.2004.

Hemmat-Jou, M. H., Safari-Sinegani, A. A., Mirzaie-Asl, A., & Tahmourespour, A., (2018). Analysis of microbial communities in heavy metals-contaminated soils using the metagenomic approach. *Ecotoxicology*, *27*(9), 1281–1291. https://doi.org/10.1007/S10646-018-1981-X.

Huse, S. M., Mark, W. D. B., Voorhis, A., Shipunova, A., Morrison, H. G., Eren, A. M., & Sogin, M. L., (2014). VAMPS: A website for visualization and analysis of microbial population structures. *BMC Bioinformatics*, *15*, 41. https://doi.org/10.1186/1471-2105-15-41.

Jacoby, R., Peukert, M., Succurro, A., Koprivova, A., & Kopriva, S., (2017). The role of soil microorganisms in plant mineral nutrition—Current knowledge and future directions. *Frontiers in Plant Science*, *8*, 1617. https://doi.org/10.3389/FPLS.2017.01617/BIBTEX.

Jat, S. L., Suby, S. B., Parihar, C. M., Gambhir, G., Kumar, N., & Rakshit, S., (2021). Microbiome for sustainable agriculture: A review with special reference to the corn production system. *Archives of Microbiology*, *203*(6), 2771–2793. https://doi.org/10.1007/s00203-021-02320-8.

Ji, X., Yao, X., Li, X., Zhu, L., Wang, J., & Wang, J., (2023). Responses of soil microbial community to herbicide atrazine contamination. *Water, Air, & Soil Pollution*, *234*(4), 1–13. https://doi.org/10.1007/S11270-023-06284-X.

Khan, N., Martínez-Hidalgo, P., Ice, T. A., Maymon, M., Humm, E. A., Nejat, N., Sanders, E. R., et al., (2018). Antifungal activity of bacillus species against *fusarium* and analysis of the potential mechanisms used in biocontrol. *Frontiers in Microbiology*, *9*, 2363. https://doi.org/10.3389/FMICB.2018.02363/BIBTEX.

Kuczynski, J., Stombaugh, J., Walters, W. A., González, A., Caporaso, J. G., & Knight, R., (2012). Using QIIME to analyze 16s rRNA gene sequences from microbial communities. *Current Protocols in Microbiology*, *SUPPL.27*. https://doi.org/10.1002/9780471729259.MC01E05S27.

Kumar, A., & Dubey, A., (2020). Rhizosphere microbiome: Engineering bacterial competitiveness for enhancing crop production. In: *Journal of Advanced Research* (Vol. 24, pp. 337–352). Cairo University. https://doi.org/10.1016/j.jare.2020.04.014.

Lakshmanan, V., Selvaraj, G., & Bais, H. P., (2014). Functional soil microbiome: Belowground solutions to an aboveground problem. *Plant Physiology*, *166*(2), 689–700. https://doi.org/10.1104/pp.114.245811.

López-García, A., Pineda-Quiroga, C., Atxaerandio, R., Pérez, A., Hernández, I., García-Rodríguez, A., & González-Recio, O., (2018). Comparison of mothur and QIIME for the analysis of rumen microbiota composition based on 16S rRNA amplicon sequences. *Frontiers in Microbiology*, *9*(DEC). https://doi.org/10.3389/FMICB.2018.03010/FULL.

Madeira, F., Pearce, M., Tivey, A. R. N., Basutkar, P., Lee, J., Edbali, O., Madhusoodanan, N., et al., (2022). Search and sequence analysis tools services from EMBL-EBI in 2022. *Nucleic Acids Research*, *50*(W1), W276–W279. https://doi.org/10.1093/NAR/GKAC240.

Malla, M. A., Dubey, A., Kumar, A., & Yadav, S., (2022). Metagenomic analysis displays the potential predictive biodegradation pathways of the persistent pesticides in agricultural soil with a long record of pesticide usage. *Microbiological Research*, 127081.

Malla, M. A., Dubey, A., Raj, A., Kumar, A., Upadhyay, N., & Yadav, S., (2022). Emerging frontiers in microbe-mediated pesticide remediation: Unveiling role of omics and in silico approaches in engineered environment. *Environmental Pollution, 299*, 118851. https://doi.org/10.1016/J.ENVPOL.2022.118851.

Malla, M. A., Dubey, A., Yadav, S., Kumar, A., Hashem, A., & Abd-Allah, E. F., (2018). Understanding and designing the strategies for the microbe-mediated remediation of environmental contaminants using omics approaches. *Frontiers in Microbiology, 9*. https://doi.org/10.3389/fmicb.2018.01132.

Mansouriieh, N., Sohrabi, M. R., & Khosravi, M., (2019). Optimization of profenofos organophosphorus pesticide degradation by zero-valent bimetallic nanoparticles using response surface methodology. *Arabian Journal of Chemistry, 12*(8), 2524–2532. https://doi.org/10.1016/J.ARABJC.2015.04.009.

Mendes, L. W., & Tsai, S. M., (2018). Distinct taxonomic and functional composition of soil microbiomes along the gradient forest-restinga-mangrove in southeastern Brazil. *Antonie van Leeuwenhoek, International Journal of General and Molecular Microbiology, 111*(1), 101–114. https://doi.org/10.1007/S10482-017-0931-6/TABLES/1.

Miranda, V., Silva-Castro, G. A., Ruiz-Lozano, J. M., Fracchia, S., & García-Romera, I., (2023). Fungal endophytes enhance wheat and tomato drought tolerance in terms of plant growth and biochemical parameters. *Journal of Fungi, 9*(3), 384. https://doi.org/10.3390/JOF9030384.

Misra, B. B., Langefeld, C., Olivier, M., & Cox, L. A., (2019). Integrated omics: Tools, advances, and future approaches. *Journal of Molecular Endocrinology, 62*(1), R21–R45. https://doi.org/10.1530/JME-18-0055.

Muccee, F., & Ejaz, S., (2020). Whole genome shotgun sequencing of POPs degrading bacterial community dwelling tannery effluents and petrol contaminated soil. *Microbiological Research, 238*. https://doi.org/10.1016/J.MICRES.2020.126504.

Munir, N., Hanif, M., Abideen, Z., Sohail, M., El-Keblawy, A., Radicetti, E., Mancinelli, R., & Haider, G., (2022). Mechanisms and strategies of plant microbiome interactions to mitigate abiotic stresses. *Agronomy, 12*(9), 2069. https://doi.org/10.3390/AGRONOMY12092069.

Narayanasamy, S., Muller, E. E., Sheik, A. R., & Wilmes, P. (2015). Integrated omics for the identification of key functionalities in biological wastewater treatment microbial communities. *Microbial Biotechnology, 8*(3), 363–368. https://doi.org/10.1111/1751-7915.12255.

Naz, M., Shah, T., Battaglia, M., Islam, M. S., Hossain, A., Iqbal, M. A., Ihsan, M. Z., et al., (2022). Insights into potential roles of plants as natural radioprotectants and amelioration of radiations induced harmful impacts on human health. *Managing Plant Production Under Changing Environment*, 311–325. https://doi.org/10.1007/978-981-16-5059-8_12.

Oliveira, S. G., Nishiyama, R. R., Trigo, C. A. C., Mattos-Guaraldi, A. L., Dávila, A. M. R., Jardim, R., & Aguiar, F. H. B., (2021). Core of the saliva microbiome: An analysis of the MG-RAST data. *BMC Oral Health, 21*(1). https://doi.org/10.1186/S12903-021-01719-5.

Phulpoto, A. H., Maitlo, M. A., & Kanhar, N. A., (2020). Culture-dependent to culture-independent approaches for the bioremediation of paints: A review. *International Journal of Environmental Science and Technology, 18*(1), 241–262. https://doi.org/10.1007/S13762-020-02801-1.

Pieterse, C. M. J., Zamioudis, C., Berendsen, R. L., Weller, D. M., Van, W. S. C. M., & Bakker, P. A. H. M., (2014). *Induced Systemic Resistance by Beneficial Microbes. 52*, 347–375. https://doi.org/10.1146/ANNUREV-PHYTO-082712-102340.

Portillo, M. C., Villahermosa, D., Corzo, A., & Gonzalez, J. M., (2011). Microbial community fingerprinting by differential display-denaturing gradient gel electrophoresis. *Applied and Environmental Microbiology*, *77*(1), 351–354. https://doi.org/10.1128/AEM.01316-10/ASSET/9806F0FB-E26F-4FB7-ACBB-A77EA39263E3/ASSETS/GRAPHIC/ZAM9991016700002.JPEG.

Poudel, M., Mendes, R., Costa, L. A. S., Bueno, C. G., Meng, Y., Folimonova, S. Y., Garrett, K. A., & Martins, S. J., (2021). The role of plant-associated bacteria, fungi, and viruses in drought stress mitigation. *Frontiers in Microbiology*, *12*, 3058. https://doi.org/10.3389/FMICB.2021.743512.

Raj, A., & Kumar, A., (2022). Integrated omics approaches to understand and improve wastewater remediation. *Omics for Environmental Engineering and Microbiology Systems*, 113–142. https://doi.org/10.1201/9781003247883-6.

Raj, A., & Kumar, A., (2022). Recent advances in assessment methods and mechanism of microbe-mediated chlorpyrifos remediation. *Environmental Research*, *214*, 114011. https://doi.org/10.1016/J.ENVRES.2022.114011.

Raj, A., Dubey, A., Malla, M. A., & Kumar, A., (2023). Pesticide pestilence: Global scenario and recent advances in detection and degradation methods. *Journal of Environmental Management*, *338*, 117680. https://doi.org/10.1016/J.JENVMAN.2023.117680.

Raj, A., Kumar, A., & Dames, J. F., (2021). Tapping the role of microbial biosurfactants in pesticide remediation: An eco-friendly approach for environmental sustainability. *Frontiers in Microbiology*, *12*, 3886. https://doi.org/10.3389/FMICB.2021.791723/BIBTEX.

Rascovan, N., Carbonetto, B., Perrig, D., Díaz, M., Canciani, W., Abalo, M., Alloati, J., et al., (2016). Integrated analysis of root microbiomes of soybean and wheat from agricultural fields. *Scientific Reports*. https://doi.org/10.1038/srep28084.

Ritchie, N. J., Schutter, M. E., Dick, R. P., & Myrold, D. D., (2000). Use of length heterogeneity PCR and fatty acid methyl ester profiles to characterize microbial communities in soil. *Applied and Environmental Microbiology*, *66*(4), 1668–1675. https://doi.org/10.1128/AEM.66.4.1668-1675.2000/ASSET/AA562BC0-8969-4B70-B75A-AA5DB35A4924/ASSETS/GRAPHIC/AM0401491004.JPEG.

Rodríguez, E., García-Encina, P. A., Stams, A. J. M., Maphosa, F., & Sousa, D. Z., (2015). Meta-omics approaches to understand and improve wastewater treatment systems. In: *Reviews in Environmental Science and Biotechnology* (Vol. 14, No. 3, pp. 385–406). Kluwer Academic Publishers. https://doi.org/10.1007/s11157-015-9370-x.

Roux, S., Faubladier, M., Mahul, A., Paulhe, N., Bernard, A., Debroas, D., & Enault, F., (2011). Metavir: A web server dedicated to virome analysis. *Bioinformatics*, *27*(21), 3074–3075. https://doi.org/10.1093/BIOINFORMATICS/BTR519.

Saad, M. M., Eida, A. A., Hirt, H., & Doerner, P., (2020). Tailoring plant-associated microbial inoculants in agriculture: A roadmap for successful application. *Journal of Experimental Botany*, *71*(13), 3878–3901. https://doi.org/10.1093/jxb/eraa111.

Saleem, M., Meckes, N., Pervaiz, Z. H., & Traw, M. B., (2017). Microbial interactions in the phyllosphere increase plant performance under herbivore biotic stress. *Frontiers in Microbiology*, *8*, 41. https://doi.org/10.3389/FMICB.2017.00041/BIBTEX.

Santos, L. F., & Olivares, F. L., (2021). Plant microbiome structure and benefits for sustainable agriculture. *Current Plant Biology*, *26*, 100198. https://doi.org/10.1016/J.CPB.2021.100198.

Schloss, P. D., Westcott, S. L., Ryabin, T., Hall, J. R., Hartmann, M., Hollister, E. B., Lesniewski, R. A., et al., (2009). Introducing mothur: Open-source, platform-independent, community-supported software for describing and comparing microbial communities. *Applied and Environmental Microbiology*, *75*(23), 7537–7541. https://doi.org/10.1128/AEM.01541-09.

Sharma, P., Singh, S. P., Iqbal, H. M. N., & Tong, Y. W., (2022). Omics approaches in bioremediation of environmental contaminants: An integrated approach for environmental safety and sustainability. *Environmental Research*, *211*. https://doi.org/10.1016/J.ENVRES.2022.113102.

Sharma, R., Shrivas, V. L., & Sharma, S., (2021). Effect of substitution of chemical fertilizer by bioinoculants on plant performance and rhizospheric bacterial community: Case study with *Cajanus cajan*. *Brazilian Journal of Microbiology*, *52*(1), 373–386. https://doi.org/10.1007/S42770-020-00418-7/METRICS.

Silva, C. C., Hayden, H., Sawbridge, T., Mele, P., De Paula, S. O., Silva, L. C. F., Vidigal, P. M. P., et al., (2013). Identification of genes and pathways related to phenol degradation in metagenomic libraries from petroleum refinery wastewater. *PLoS One*, *8*(4). https://doi.org/10.1371/JOURNAL.PONE.0061811.

Singh, T., & Singh, D. K. (2017). Phytoremediation of organochlorine pesticides: Concept, method, and recent developments. *International Journal of Phytoremediation*, *19*(9), 834–843. https://doi.org/10.1080/15226514.2017.1290579.

Timmusk, S., Grantcharova, N., & Wagner, E. G. H., (2005). *Paenibacillus polymyxa* invades plant roots and forms biofilms. *Applied and Environmental Microbiology*, *71*(11), 7292–7300. https://doi.org/10.1128/AEM.71.11.7292-7300.2005/ASSET/CC81B596-5663-4511-9EE2-D3AC29ECA32A/ASSETS/GRAPHIC/ZAM0110559620005.JPEG.

Tkacz, A., & Poole, P., (2021). The plant microbiome: The dark and dirty secrets of plant growth. *Plants People Planet*, *3*(2), 124–129. https://doi.org/10.1002/ppp3.10167.

Van, D. H. E., Geertz-hansen, H. M., Genee, H. J., Malla, S., Otto, M., & Sommer, A., (2015). deFUME: Dynamic exploration of functional metagenomic sequencing data findings background. *BMC Research Notes*, *8*. https://doi.org/10.1186/s13104-015-1281-y.

Walvekar, V. A., Bajaj, S., Singh, D. K., & Sharma, S., (2017). Ecotoxicological assessment of pesticides and their combination on rhizospheric microbial community structure and function of *Vigna radiata*. *Environmental Science and Pollution Research*, *24*(20), 17175–17186. https://doi.org/10.1007/s11356-017-9284-y.

Wang, Z., Zhang, X. X., Lu, X., Liu, B., Li, Y., Long, C., & Li, A., (2014). Abundance and diversity of bacterial nitrifiers and denitrifiers and their functional genes in tannery wastewater treatment plants revealed by high-throughput sequencing. *PLoS One*, *9*(11). https://doi.org/10.1371/JOURNAL.PONE.0113603.

White, J. F., Kingsley, K. L., Zhang, Q., Verma, R., Obi, N., Dvinskikh, S., Elmore, M. T., et al., (2019). Review: Endophytic microbes and their potential applications in crop management. *Pest Management Science*, *75*(10), 2558–2565. https://doi.org/10.1002/PS.5527.

Wolinska, A., Kuzniar, A., & Galazka, A., (2020). Biodiversity in the rhizosphere of selected winter wheat (*Triticum aestivum* L.) cultivars-genetic and catabolic fingerprinting. *Agronomy*. https://doi.org/10.3390/agronomy10070953.

Wu, S., Zhu, Z., Fu, L., Niu, B., & Li, W., (2011). WebMGA: A customizable web server for fast metagenomic sequence analysis. *BMC Genomics*, *12*. https://doi.org/10.1186/1471-2164-12-444.

Yergeau, E., Bell, T. H., Champagne, J., Maynard, C., Tardif, S., Tremblay, J., & Greer, C. W., (2015). Transplanting soil microbiomes leads to lasting effects on willow growth, but not on the rhizosphere microbiome. In: *Frontiers in Microbiology* (Vol. 6, p. 1436). https://doi.org/10.3389/fmicb.2015.01436.

Yoo, K., Lee, T. K., Choi, E. J., Yang, J., Shukla, S. K., Hwang, S., & Park, J., (2017). Molecular approaches for the detection and monitoring of microbial communities in bioaerosols: A review. *Journal of Environmental Sciences, 51*, 234–247. https://doi.org/10.1016/J.JES.2016.07.002.

Yu, K., & Zhang, T., (2012). Metagenomic and metatranscriptomic analysis of microbial community structure and gene expression of activated sludge. *PLoS One*, *7*(5). https://doi.org/10.1371/JOURNAL.PONE.0038183.

Zaheer, M. S., Ali, H. H., Iqbal, M. A., Erinle, K. O., Javed, T., Iqbal, J., Hashmi, M. I. U., et al., (2022). Cytokinin production by *Azospirillum brasilense* contributes to increase in growth, yield, antioxidant, and physiological systems of wheat (*Triticum aestivum* L.). *Frontiers in Microbiology*, *13*, 1613. https://doi.org/10.3389/FMICB.2022.886041/BIBTEX.

CHAPTER 2

Microbial Endophytes as Warriors for Plant Defense Against Abiotic and Biotic Stress

ANAMIKA DUBEY,[1] DIKSHA SAIYAM,[1] and ASHWANI KUMAR[1,2]

[1]*Metagenomics and Secretomics Research Laboratory, Department of Botany, Dr. Harisingh Gour Vishwavidyalaya (A Central University), Sagar, Madhya Pradesh, India*

[2]*Metagenomics and Secretomics Research Laboratory, Department of Botany, University of Allahabad (A Central University), Prayagraj, Uttar Pradesh, India*

ABSTRACT

The most exciting and fascinating scientific advances in the past few years have been the realization and understanding that the diverse and enormously active endophytic bacterial communities are not only the 'passengers' of their host plants but that instead, these microbial communities play a key and imperative role in plant life. Current studies provide evidence that these microbial communities, besides serving several different host functions such as growth, health, development, and resistance to different stresses, also improve the capacity of taking up nutrients via the acquisition of nutrients from the soil. Understanding the interactions between endophytes and host plants requires the modern-day molecular approaches of high throughput detection. However, the modern "omics" approaches appear to be useful in unraveling the gray areas of endophytism. This chapter highlights the role of endophytic bacteria and their multifarious interactions that are beneficial for plant growth and health.

Advances in Plant Microbiome Research for Climate Resilient Agriculture: Toward Sustainable Farming. Ashwani Kumar (Ed.)

2.1 INTRODUCTION

The concept of microbiome research and its significance for the host organism have been the focus of present-day research, as it has led to noteworthy and significant advances in understanding the enormous power of microbes (Hadrich, 2018). Every dynamic biogeochemical cycle on Earth is driven by microorganisms, and this global nutrient cycling has a direct impact on productivity and ecosystem health (Malla, Dubey, Kumar et al., 2019; van der Heijden & Hartmann, 2016). Since being introduced to the earth, terrestrial plants have developed symbiotic relationships with various microbial groups (Field et al., 2015). The plant microbiome research predates their human, and animal counterparts and is recognized as an important treasure trove of microbial diversity because of their crucial role in fixing nutrients for the plant, especially phosphorus and nitrogen (Hunter, 2016). According to Sprent (2008), the plant-endophytic connection is thought to have developed around 60,000 years ago and has been important to terrestrial ecosystems, benefiting both partners. However, studies on plant-microbiome research have completely changed the scenario and shown that these microbial communities are for more flexible and versatile and contain genes coding for the novel traits beneficial to host organisms (Ali et al., 2014). By defending the host plant against various stress circumstances (abiotic and biotic stressors), microbial endophytes play a crucial role in promoting the health and fitness of their host plant. They also help in enhancing plant growth and production (Lata et al., 2018; Lugtenberg et al., 2016). This co-evolution must be examined in order to identify the causes influencing the coexistence of both companions and monitor the benefits resulting from this interaction because of their essential contribution to the host organism (Turner et al., 2013). Induction of plant genes expressions and signaling pathways, in the presence of microbial endophytes provides evidence about their effects on host plant (Berendsen et al., 2015). The modern "omics" approaches in combination with different computational approaches, may provide a comprehensive facet of the endophytic lifestyle (Dubey et al., 2020, 2021; Kaul et al., 2016). In spite of the enormous progresses in the plant-microbiome research, a major challenge ahead for the future research is to link the microbiome composition to plant functioning. Can the plant microbiome be altered, for instance, to increase agricultural output and environmental sustainability? In order to address the mechanisms leading to the community assemblage and its functioning in and on plants, additional basic and applied research are needed. Deciphering the interactions between and between the many microbial species that colonize roots, stems, petioles, and leaves is also necessary in order to comprehend and investigate the control of the endophytic microbiome. A number of regulating factors, including interactions between microbes, environmental conditions, host genotype, host immune system, nutrition, and soil quality, are some of the primary factors influencing the establishment and structure of communities (Edwards et al., 2015; Hacquard et al., 2015). The main environmental factors influencing the makeup of the microbial community are pH, physiological factors, temperature, and oxygen availability (Carmody et al., 2015; Hacquard et al., 2015). The aim of the present chapter is to provide a detailed picture of the role of endophytic microbes on host growth and health. Additionally, metabolomics of endophytes along with the community structure and application omics tools to understand the plant-endophyte interactions are also discussed. A complete overview of this whole process is given in Figure 2.1.

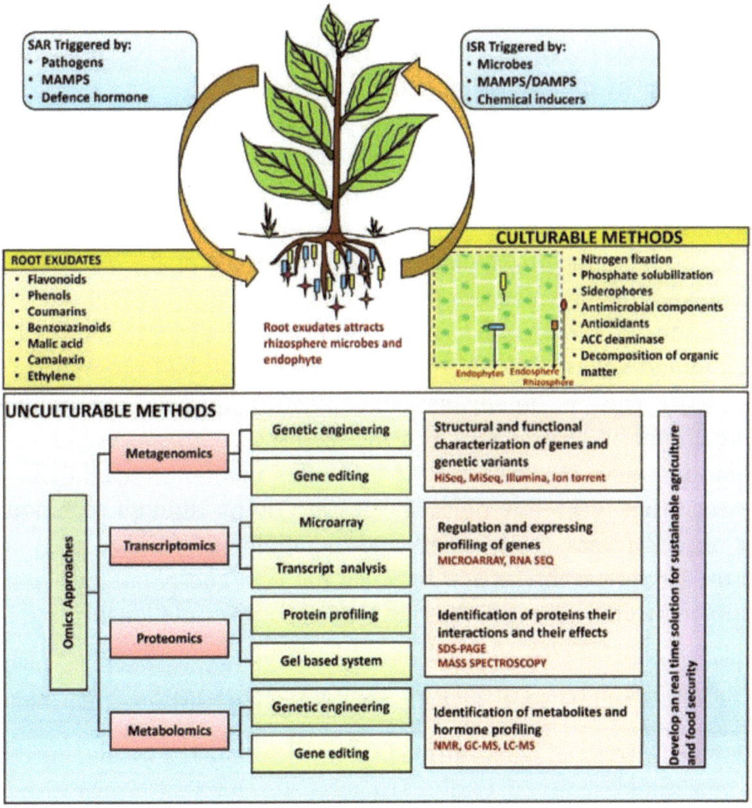

FIGURE 2.1 Diagrammatic flow of strategies being developed to investigate the plant microbiome from various compartments and to apply the omics methodology to identify the cultivable and uncultivable microbiome for promoting plant growth.

2.2 COMMUNITY STRUCTURE OF THE ROOT MICROBIOME: WHO ARE THEY (PHYLOGENETIC AFFILIATIONS)?

The host-associated microbiome is mostly dominated by a small number of phyla, despite the great majority of microbial variety found in the biosphere (now, less than 80 bacterial phyla are identified). Different plant species' root endophytic compartments and rhizosphere microbiomes are frequently colonized by bacteria from the three main phyla Proteobacteria, Actinobacteria, and Bacteroidetes (Hacquard et al., 2015). However, many Acidobacteria bacteria are not present in the endophytic compartment (Bulgarelli et al., 2015). In contrast to the surrounding soil, microbial taxa belonging to phylum Proteobacteria are consistently enriched in endosphere/rhizosphere parts of the annual and perennial plants (Edwards et al., 2015; Ofek-Lalzar et al., 2014; Peiffer et al., 2013; Schlaeppi et al., 2013; Shakya et al., 2013; Zarraonaindia et al., 2015). This presence of certain phyla within or outside the roots reflects the niche adaptation and ability of the bacterial communities to efficiently invade different plant species.

While, culture-independent surveys, such as amplicon (16S rRNA) based sequencing approaches attempt to answer the diversity (who's there?), within a particular microbial community, shotgun metagenomic sequencing on the other hand attempts to reveal the functional core diversity "(what can they do?)" of the microbial communities (Malla et al., 2018). Moreover, the functional, approaches are used to enumerate and unravel the structural composition and functional diversity of the microbial communities and have helped to unravel more complex and intricate microbial ecosystems (David et al., 2014). The host organisms (either plants or animals) and microbes rely on one another: the host provides the microbes with space and resources, and the microbial communities in turn offer function essential for health. Knowing what these microbial species are doing through techniques like metatranscriptomics, proteomics, and metabolomics will help us comprehend the symbiotic interaction between the host and plant and identify the common expected activities that are crucial for this relationship.

2.3 EXPLORATION OF THE PLANT MICROBIOME

The roots of terrestrial plants thrive in soil, one of the richest and most diverse microbial ecosystems on Earth (Tringe et al., 2005). Comparative studies performed by several researchers revealed that soil properties and nutrient availability are the primary determinants of root microbiome and consequently, the endophytic microbial communities (Agler et al., 2016; Schreiter et al., 2014). Just as, the gut microbiome interacts with the food consumed by the animals, the root microbial communities mediate the soil-based food of the plants (Ramírez-Puebla et al., 2013). Similar to host-microbe interactions across the animal kingdom, the plant is also compartmentalized by the diverse microbial taxa (Hacquard et al., 2015; Kostic et al., 2013). Whilst the category of the microbial communities that frame plant microbiomes are largely conserved, much of the variations do exist in the species composition of these microbial communities across the host organisms. One of the key factors in determining the maintenance and the microbial community's composition depends on the release of organic molecules by plants *via* rhizodeposition processes. This process of releasing the organic molecules into the rhizosphere influences the physical and chemical composition of the plant rhizosphere and provides organic matter and signaling molecules for microbial growth. Recent findings in plant microbiome research stated that plant microbiomes can promote growth, facilitate the efficiency of nutrient uptake and their use, and buffer the plant hosts against different stress conditions. Several previous studies conducted by different scientists suggested that despite the presence of pathogens, infectious and unfavorable growth conditions, the plants in some areas produces are less susceptible to disease than that of other areas, the reason being that soils in these areas support the plant growth and health via microbiome (Berendsen et al., 2018; Martínez-Hidalgo et al., 2018; Mendes et al., 2018; Mendes et al., 2011).

Notably, the soil microbial communities can proliferate inside the roots as advantageous endophytes and control the growth and development of host plants Hardoim et al. (2008), with effects ranging from enhanced plant health (Waghunde et al., 2017) to disease resistance (Kane, 2011; Raghavendra & Newcombe, 2013) and from crop production (Patle et al., 2018) to phytoremediation (Li et al., 2012; Mesa et al., 2017).

2.4 ENHANCING ABIOTIC AND BIOTIC STRESS RESISTANCE IN PLANTS BY ENDOPHYTES

2.4.1 INCREASE IN RESISTANCE TO ABIOTIC STRESSES

It is thought that the functional consistency of the microbial community patterns supports the idea of underlying principles and the factors that govern community composition and formation. Although the initial microbiome structure and makeup are comparable to those of their respective seed banks, including air and soil, they become more and more plant-specific as the plant grows and develops (Copeland et al., 2015; Shi et al., 2015; Sugiyama et al.,

2014). Many factors, including abiotic stresses like oxidative stress, salinity, alkalinity, drought stress, extreme temperatures, and heavy metal toxicity are threats to agricultural ecosystems (Dubey et al., 2019; Malla et al., 2018). Endophytes adopt various molecular mechanisms to mitigate the effect of stress. Stress tolerance in plants is mainly induced by generating free radicals such as ROS, synthesis of antistress metabolites, and via expression of stress-responsive genes (Lata et al., 2018). Phytohormones such as auxin, gibberellic acid, abscisic acid (ABA), cytokinin, Jasmonic acid (JA), ethylene, and salicylic acid (SA) play key roles in abiotic stress tolerance in plants (Wani et al., 2016). Waqas et al. (2012) in their study, reviewed that ABA-mediated and stomatal motions interact with one another. The presence of helpful microbes inhabiting the plant endosphere modifies the ABA production and ABA-mediated signaling pathways, and these microbes' mediated pathways support plant development under salt-stress conditions. Upregulation of different genes including aquaporins and dehydrins plays a role in mitigating the abiotic stresses in rice plants when inoculated by endophytic *Trichoderma harzianum* (Pandey et al., 2016).

Some types of plants can thrive on polluted soils. The plants have some genetic and physiological defenses against various soil contaminants, including those brought on by anthropocentric activity. The recruitment of bacterial endophytes is an important plant process involved in phytoremediation. More significantly, certain bacterial endophytes that can withstand or resist high concentrations of contaminants also have characteristics that encourage plant development. In a recent study, several bacterial endophytes were isolated from the Zn/Cd hyperaccumulator plant *Sedum plumbizincicola* and five isolates exhibiting PGPB activities were selected for further study (Ma et al., 2015; Ullah et al., 2015). These activities included ACC deaminase activity, production of indole-3-acetic acid, synthesis of siderophores, and solubilization of phosphorus. In addition, the strains showed high resistance to growth inhibition by heavy metals including Cd, Zn, and Pb. The effects of inoculating these metal-resistant ACC-utilizing strains on the growth of *S. plumbizincicola* and its uptake of Cd, Zn, and Pb in multi-metal contaminated soils were also evaluated in pot experiments. One particular strain, *Bacillus pumilus* strain E2S2 significantly increased the plant's Cd uptake, and increased the plant root and shoot length, as well as fresh and dry biomass compared with non-inoculated plants. Thus, endophytic bacteria from *S. plumbizincicola* improved its phytoextraction capacity and at the same time, promoted plant growth.

The study conducted by Kong et al. (2015) showed the effect of *Sinorhizobium* meliloti strain CCNWSX0020 the rhizobial endosymbiont on the host plant Medicago *lupulina* under copper stress. This particular strain enhances nitrogen content as well as the growth of the plant. Additionally, the rhizobial symbiosis also promotes the accumulation of copper in the roots and shoots of the plant. Therefore, many genes which are involved in antioxidant responses were also upregulated in plants when treated with the endophytic bacterium in the presence of high levels of copper. Hence, *S. meliloti* not only promotes metal uptake and plant growth but also induces the plant's antioxidative defense responses under copper stress. Yu et al. (2022) studied that; Plant-specific proteins called BURP proteins first arose as plants migrated from aquatic to terrestrial habitats. Rice transgenics with ectopic expression of PpBURP2 were treated with various abiotic stress treatments to learn more about the role of BURP protein in moss. According to the findings, PpBURP2 was expressed ectopically, which improved the plant's ability to withstand drought stress in adulthood as well as osmotic and salt stressors in seedlings. Additional PpBURP2 expression increased Cd2+ tolerance and decreased Cd2+ buildup in rice leaves. Jin et al. (2022) in their study, examined how the soybean DNAJ gene GmDNAJC7 affects abiotic stress tolerance in plants. Treatments of soybean roots with alkaline salt, salt, and drought enhanced the expression of the GmDNAJC7 gene, indicating a potential involvement of this gene in the response of soybeans to these stressors.

2.4.2 INCREASE IN RESISTANCE TO BIOTIC STRESSES

Endophytes are a significant class of ubiquitous, fascinating, and diverse plant symbionts that live in the host tissues asymptomatically without manifesting any indications of disease (Cui et al., 2017; Porras-Alfaro & Bayman, 2011a). These symbionts reside in the internal tissues of their host plants *via* quiescent infections (Card et al., 2016; Jia et al., 2016), and have the potential to provide defense against a variety of pathogens *via* elicitation of different plant responses (Khare et al., 2018; Pappas et al., 2018). Endophytic microbes can cope with various Phyto-pathogens by inducing ISR against them (Sahu et al., 2017). Many foliar endophytic microbes are well known to regulate the host gene expression, affecting the defensive pathways and physiological responses of the host plants (Salam et al., 2017). SA and JA, two different phytohormones, are known to play an important role in pathogen attacks *via* the activation of plant defense systems (Xu et al.,

2018). Endophytes enhance plant growth, *via* the production of a variety of hormones and other regulatory chemicals that increase plant accessibility to nutrients, such as nitrogen, potassium, and phosphorus. Gilbert et al. (2018); and Waqas et al. (2015) in a study, reported that gibberellin-producing endophytes are known to increase the host's resistance against the intruders through JA and SA pathways. Bacterial endophytes are reported to synthesize a variety of novel organic compounds and antifungal metabolites, with a broad spectrum of antimicrobial activity against different phytopathogenic fungi, bacteria, and nematodes. A recent study by Sheoran et al. (2015) described that the endophytic bacteria *Pseudomonas putida* BP25 associated with black pepper suppresses the growth of various phytopathogens such as *Pythium myriotylum, Phytophthora capsici, Athelia rolfsii, Rhizoctonia solani, Gibberella moniliformis, Colletotrichum gloeosporioides*, and parasitic nematodes, like *Radopholus similis*, by producing a variety of volatile substances. For the very first time, Etminani & Harighi (2018) in his study reported the antagonistic strains of endophytic bacteria *Bacillus, Pseudomonas, Pantoea, Stenotrophomonas,* and *Serratia*, isolated from wild pistachio showing positive biocontrol of phytopathogenic fungus *Pseudomonas tolaasii and Pseudomonas syringae*. Similarly, Hashem et al. (2017) in their study reported the biocontrol of *Macrophomina phaseolina* causing (Tassi) goid disease in mung bean by using *Bacillus subtilis* (Bera 71) as a biocontrol agent. A similar study concluded by Tsalgatidou et al. (2022) in this study, they introduced the novel Bacillus halotolerans Cal.l.30 strain, endowed with a rich array of secondary metabolite biosynthetic gene clusters. A large repertoire of CAZyme genes involved in antifungal activity and equipped with the capacity to produce and secrete secondary metabolites (SMs) and other novel compounds that are involved in antibiosis and/or host plant systemic resistance induction and the successful colonization of host plant tissues. In another similar study reported by Lastochkina et al. (2022) primary goal of this study was to assess the impact of the endophytic bacteria, *Bacillus subtilis* (strain 10-4) or its combination with SA on a few resistance and quality parameters of stored *P. infestans* infected potato tubers.

2.5 MODULATION OF PLANT'S IMMUNE SYSTEM BY ENDOPHYTES

At present, one of the biggest challenge challenges for the researcher is to find and track the mechanism, of how the endophytic microbes flourish inside the host plant. Plants usually establish interactions with microbes that have detrimental or beneficial effects on health and fitness. It has been stated that symbiosis with endophytic microbes confers additional defensive mechanisms to the host plants (Cameron et al., 2013; Pieterse et al., 2014a; Porras-Alfaro & Bayman, 2011b; Pozo et al., 2015a; Pozo & Azcón-Aguilar, 2007). More precisely, these symbiotic organisms modulate the plant hormonal pathways implicated in defenses, colonize the host plant and establish symbiosis (Gutjahr, 2014; Pozo et al., 2015b). Endophytes are an active source of bioactive compounds such as alkaloids, these compounds are well known to confer protection to the host plant (Schardl et al., 2013). The association between the plant and microbes can have significant consequences on the host's immune system (Balmer et al., 2015; Pieterse et al., 2014b). Some species of mycorrhizal fungi and rhizobacteria induce certain types of systemic defenses, known as MIR (mycorrhiza-induced resistance) and ISR (induced systemic resistance), respectively (Pieterse et al., 2014b; Pozo et al., 2015b; Van Der Ent et al., 2009). Table 2.1 shows the diverse role played by endophytes.

Piercing through to the host, endophytes have to cross the first line of highly sophisticated defense of the plant immune system, this process involves the identification of conserved molecules, termed pathogen – or microbe-associated molecular patterns (PAMPs or MAMPs) by the host plants (Newman et al., 2013). Peptidoglycans, lipopolysaccharides (LPS), elongation factor TU (EF-Tu; elf18/26), flagellin (Flg; flg22), bacterial superoxide dismutase (SOD), β-Glycan bacterial cold shock proteins (RNP1motif), β-glucans from oomycetes, and chitin are the most worked upon MAMPs (Newman et al., 2013). These MAMPs are recognized by specific receptor molecules present on the surface of plant cells and are termed pattern recognition receptors (PRRs). Similar to microbial endophytes, fungal endophytes, also have specific receptors, the chitin-receptors (PR-3) recognize chitin oligomers formed on the fungal cell wall (Sánchez-Vallet et al., 2015).

Endophytes are also known, to work out mechanisms to protect themselves from plant defense mechanisms. For example, Cord-Landwehr et al. (2016) in their study carried stated that fungal endophytes either mask the chitin by deacetylating it into chitosan *via* chitin deacetylases enzyme, or by covering it with other molecules, hence, preventing themselves from being recognized. Studies also suggest that endophytic bacteria produce their MAMPs, which are either not recognized by PRRs present on the cell surface of the host plants or plants prompt a weak defensive response compared to pathogenic interactions (Vandenkoornhuyse et al., 2015). Similarly, Trdá et al. (2015) in their study concluded that grapevine exhibited a different type of perception in response to flagellin (FLS 2) when encountered with

TABLE 2.1 Diverse Roles Played by Endophytes in Association with Host Plants

SL. No.	Endophytes	Host Plant	Source	Reported Roles	References
1.	*Variovorax paradoxus* S110^T and *Variovorax gossypii* JM-310^T	Alfalfa	Nodules	Improves the adaptation of legumes to degraded soils in soil-recovery programs.	Singh et al. (2022)
2.	*Bacillus amyloliquefaciens* HAPH2, *Priestia endophytica* HAPH5, *Bacillus subtilis* HAPH7, *Bacillus toyonensis* HAPH8, *Halomonas sulfidaeris* HAPH11, *Isoptericola halotolerans* HAPH12, *Planomicrobium* soli HAPH15, and *Pseudomonas kilonensis* HAPH16	Black saxaul (*Haloxylon aphyllum Minkw.*)	Roots and shoots	Enhancing plant growth by showing plant growth-promoting traits like nitrogen fixation, phosphate solubilization, auxins, and siderophores production, showing ACC deaminase activity.	Shurigin et al. (2022)
3.	G5 (*Lelliottia amnigena*), G20 (*Alcaligenes* sp.), Y2 (*Pantoea agglomerans*), J2 (*Bacillus megaterium*)	*Pairs polyphylla var. yunnanensis* (Paris L.)	Rhizomes	Enhancing plant growth by showing plant growth promoting (PGP) traits.	Tao et al. (2022)
4.	*Priestia megaterium, Brucella haematophila and Microbacterium paraoxydans*	Chickpea	Root	Showed PGP traits and shows the antagonistic ability for *Fusarium oxysporum*	Khanna et al. (2022)
5.	*Burkholderia gladioli, Streptomyces achromogenes*, *Bacillus mojavensis* CS4EB32 and *Burkholderia gladioli* E39CS3	*Crocus sativus L.* (saffron crocus)	Corms and the shoots.	Plant growth-promoting properties and inhibitory activities against the specific fungal pathogens of saffron.	Ahmad et al. (2022)
6.	*Bacillus* spp., *Paenibacillus* spp., *Lysinibacillus* spp., *Erwinia* spp., *Herbaspirillum* spp., *Micrococcus* spp., *Methylobacterium* spp., *Sphingomonas* spp., *Bosea* spp., and *Aurantimonas* spp.	Albino tea	leaves	PGP bacteria isolated from albino leaves are expected to become a plant growth promoter in the future.	Jia et al. (2022)
7.	*Bacillus aryabhattai*	*Glyceria chinensis* (Keng)	Stems and roots	The expression of novel transcriptomics changes to stimulate plant growth.	Xu et al. (2022)

TABLE 2.1 *(Continued)*

SL. No.	Endophytes	Host Plant	Source	Reported Roles	References
8.	*Bacillus*, *Burkholderia*, *Acinetobacter*, *Pseudomonas*, *Enterobacter*, *Pantoea*, and *Paenibacillus.*	Sugarcane cultivar Yunzhe-99-91	Leaves, stems, and roots.	Enhancing plant growth by showing plant growth-promoting (PGP) traits and resistance to biotic and abiotic stresses.	Di et al. (2022)
9.	*Microbacterium proteolyticum*, *Microbacterium hydrothermale*, *Bacillus altitudinis*, and *Pseudonocardia alni*	Maize (*Zea mays*)	Stem and leaf.	Endophytes biofortify and increase the bioavailability of food crops with important micronutrients like Fe and Zn.	Verma et al. (2022)
10.	*Bacillus licheniformis* R1, *Bacillus* sp. R2, *Agrobacterium tumefaciens* R6, *Bacillus subtilis* RS3, *Bacillus subtilis* RS6, and *Lysinibacillus fusiformis* RS9	*Momordica charantia* L.	Roots	PGP traits showed the possibility to use as plant or soil inoculants to enhance growth and disease management in plants.	Singh et al. (2022)
11.	*Halomonas*, *Agrobacterium*, *Exiguobacterium*, *Jonesia*, *Stenotrophomonas*, *Pseudomonas*, *Alishewanella*, *Kosakonia*, *Bacillus*, *Paracoccus*, and *Pannonibacter*	*Sporobolus specatus* (*Vahr*) *Kunth* and *Cyperus laevigatus* L.	Roots, stems, and leaves.	Showed PGP traits and tolerance for various stress conditions.	Enquahone et al. (2022)
12.	*Bacillus cereus*, *Pseudomonas otitidis*, and *Pseudomonas* sp.	Soybean (*Glycine max*)	Roots	Enhanced drought stress tolerance by the production of osmolytes, sugars, and reduction of oxidative stress.	Dubey et al. (2021)
13.	*Pantoea vagans* IALR611, *Pseudomonas psychrotolerans* IALR632, *Bacillus subtilis* IALR1033, *Bacillus safensis* IALR1035 and *Pantoea agglomerans* IALR1325	Pepper and tomato	Root, leaf, and stem.	Enhancing plant growth by phosphate solubilization.	Mei et al. (2021)

TABLE 2.1 *(Continued)*

SL. No.	Endophytes	Host Plant	Source	Reported Roles	References
14.	*Bacillus thuringiensis* PB2 and *Brevibacillus agri* PB5 and fungus *Alternaria sorghi* PF2, *Penicillium commune* PF3	Bean (*Phaseolus vulgaris* L.)	Root	Endophytic microbes have been shown to play an important role in plant growth, development, and fitness by producing growth-promoting substances and secondary metabolites.	Ismail et al. (2021)
15.	*Bacillus subtilis* (*AR5*) and *B. thuringien sis* (*BR1*)	*Arthrocnemum macrostachyum* and *Spergularia marina*	Shoot tip and root	Bacterial strains can induce salinity tolerance, enhance nutrient uptake and promote broad bean growth under salinity stress.	Mahgoub et al. (2021)
16.	*Brevibacillus agri* Af.13, and Af.14, *Bacillus* s., Fm3, *Bacillus* sp. Fm.4 and *Bacillus* sp. Fm.6	*Fagonia mollis* Delile and *Achillea fragrantissima* (Forssk) Sch. Bip.)	Leaves	Endophytes are good candidates as plant growth-promoting inoculants to help reduce chemical input in conventional agricultural practices and increase nutrient uptake and stress resilience in plant species.	ALKahtani et al. (2020)
17.	*Pseudomonas* spp. and *Bacillus* spp.	Apple (*Malus domestica*)	Buds	Modulate cellular redox balance, shoot growth, and protein expression under *in-vitro* conditions.	Tamošiūnė et al. (2018)
18.	*Aspergillus oryzae*	Radish (*Raphanus sativus*)	*Raphanus sativus*	Showed PGP traits and provide protection against *Plutella xylostella* an herbivore pest.	Sun et al. (2018)
19.	*Bacillus subtilis* and *B. cereus*, *Penicillium crustosum* and *P. chrysogenum*	Felty germander (*Teucrium polium*)	Leaves	Enhancing plant growth by showing plant growth promoting (PGP) traits.	Hassan (2017)
20.	*Enterobacter* sp. *strain* PDN3	Populus (*Populus deltoides*)	Cuttings of populus	Phytoremediation of trichloroethylene (TCE)	Doty et al. (2017)

two different types of endophytes such as *Burkholderia phytofirmans* and bacterial pathogens *Pseudomonas aeruginosa* and *Xanthomonas campestris*. Sun et al. (2022) in their reported study findings demonstrate that fungal colonization alters plant immunology, including oxidative signaling and plant cell death, in response to environmental N circumstances, and they highlight SA signaling as a possible regulator of the relationship between rice and *Ph. liquidambaris*. In the case of the production of reactive oxygen species (ROS), as a plant defense system, endophytes defend themselves by generating different antioxidant enzymes (Staerck et al., 2017).

2.6 USE OF OMICS TOOLS FOR ENHANCED UNDERSTANDING OF PLANT–ENDOPHYTE INTERACTIONS

A comprehensive and detailed understanding of plant-endophyte interaction is important to realize the value of these microbial communities, which are the source of industrially important enzymes, novel metabolites, and stress relievers of the host plant. However, several valuable facets of endophytes are still unknown. Modern high-throughput genomic studies have revolutionized microbiome research by uncovering the gray areas of endophytism. A detailed and comprehensive knowledge of the mechanism of host infestation and the role of endophytes could be used to improve agricultural management in terms of, biocontrol, plant growth promotion, and phytoremediation. Microarray, next-gen sequencing, whole genome sequencing, comparative genomics, metagenomics, metatranscriptomics, metabolomics, and fluxomics are some of the modern molecular-based approaches harnessed to disentangle the host-endophyte relationship. Metagenomic approaches along with *in silico* analysis will provide novel phylogenetic and functional insight. Modern omics-based techniques and approaches need to be explored to study endophyte biology along with their putative role in host ecology. Metagenomics has seen more application recently as a result of its ability to qualitatively and quantitatively analyze the microbial makeup of bacteria and fungi in the plant microbiome (Kumar & Dubey, 2020; Melcher et al., 2014). Studies in metagenomics combine genomics, bioinformatics, and systems biology to investigate the microbial genomes that were directly extracted from an environmental sample. High-throughput screening techniques (HTS) and new bioinformatics tools, along with affordable next-generation sequencing (NGS) technology, have significantly advanced metagenomic research (de Medeiros Azevedo et al., 2021; Kumar et al., 2015). Amplicon and shotgun metagenomics are two separate approaches for doing metagenomic analyses (Gilbert & Dupont, 2011). The amplicon metagenomics method for identifying microbial diversity involves directly extracting DNA from an environmental sample (such as the rhizosphere, phyllosphere, or endosphere), amplifying 16S, 18S, and/or ITS genes via PCR, sequencing the amplicons, and comparing amplicon similarity to databases (such as those for ribosomal genes or ITSs) that are readily available. Shotgun metagenomics is based on sequencing a whole sample's DNA. Given the emergence of more effective NGS technology and bioinformatics tools (Abraham et al., 2020). The advantage of shotgun metagenomics is that it provides a better knowledge of the roles in addition to the taxonomic assemblage of the microbes (Enagbonma et al., 2019). For their studies, many researchers used omics approaches. Sessitsch et al. (2012) studied the first metagenomic approach to analyze an endophytic bacterial community resident inside the roots of rice, one of the most important staple foods. Babalola et al. (2020) studied, using a shotgun metagenomic technique, the dataset displays the root endophytic microbial community profile of maize (*Zea mays* L.), one of the major food crops in South Africa. To the best of our knowledge, this is the first description using a shotgun metagenomics method to illustrate the endophytic microbial diversity in maize plants. Using the NovaSeq 6,000 machine, high throughput sequencing of the community's whole DNA was completed. Another study conducted by Fadiji et al. (2021a, b) in his study, used shotgun metagenomics to examine the whole DNA from maize roots planted with various fertilizations and without fertilization to evaluate the effect of fertilizer treatments on the functional diversity of endophytic microbiomes. Das & Dhal et al. (2022) in a study, utilizing amplicon metagenomics techniques, it was attempted to examine the diversity of endogenous microbes in rice roots and their connection to soil salinity and physicochemical elements in the salt-stressed Sundarbans region of India.

While metatranscriptomics deals with the transcriptome sequencing of an environmental sample, allowing for the analysis of the expressed RNAs of the accompanying microbiota. Metatranscriptomics serves as a supplement to metagenomic data in this way, enabling the precise measurement of the transcriptional profile of the genomes that make up a microbial community subjected to a certain environmental situation (de Medeiros Azevedo et al., 2021; Marchesi & Ravel, 2015). However, the rRNA dominance in the samples from metatranscriptomics enables a thorough investigation of the complete microbiome without the need to first choose the taxonomic groups that will be included in the study. This is less difficult than sample enrichment with mRNA, which removes the need for a PCR-based phase and may be done immediately on many samples (Kumar & Dubey, 2020). Metatranscriptomics analysis of soybean plants has revealed the presence of several small RNA sequences unrelated to the soybean genome. Interestingly comparative analysis of the obtained sequences established the presence of various pathogenic, symbiotic, and free-living microbes in different samples of the soybean plant (Gomes Molina et al., 2012). Saminathan et al. (2018) studied, to comprehend the variety and purpose of microorganisms in connection to the metabolism of carbohydrates in mature watermelon fruits. They analyzed the red (PI459074, Congo, and SDRose) and yellow (PI227202, PI435990, and JBush) fruit-flesh cultivars of geographically and metabolically varied watermelon cultivars using 16S metagenomics and RNAseq metatranscriptomics. Paolinelli et al. (2022) studied the microbiota of mature field grown Vitis vinifera cv was characterized using a metatranscriptomics analysis. Malbec, whether it is symptomatic or not. This study's molecular data will be used to construct future-focused molecular quantification for certain species.

Numerous programs, such as omeSOM, PRIME Plant, and MetGenMAP, are available to manage the whole amount of data produced by various "omics" approaches. The findings of metagenomic sequencing analysis must be further analyzed since it produces enormous amounts of data (Kumar & Dubey, 2020; Milone et al., 2010). Presently developed shotgun metagenomics sequencing-based taxonomical profiling software can be separated into two distinct groups: the alignment-free software and the alignment-based. Alignment-based software, such as BLASTN (Camacho et al., 2009; Johnson et al., 2008; Xu et al., 2022.). Which aligns sequences at the nucleotide level, and Diamond, which aligns at the protein level, was thought to have high sensitivity and has been used as the standard for metagenomics profiling (Xu et al., 2022). Furthermore, quantitative insights into microbial ecology (QIIME), MEGAN, and CARMA are three of the many software programs that are available for amplicon sequencing analysis and are subsequently used for 454 ribosomal pyro-tag sequences or Sanger sequencing. These programs are crucial and frequently used for metagenomic analysis (Chaparro et al., 2014; Huson & Weber, 2013).

2.7 PLANT-MICROBIOME SEQUENCING

The innovative and ground-breaking changes in our ability to understand and explore the role of the microbiome came with the advent of high-throughput sequencing that has allowed comprehensive characterization of the plant microbiome using multi-omics approaches before their cultivation. One of the most widely accepted methods to characterize microbial communities is amplicon sequencing, using the highly informative phylogenetic marker (16S rRNA gene), a highly conserved gene in bacteria (Malla, Dubey, Yadav et al., 2019; Malla et al., 2018). However, this approach lacks species and strain level resolution as well as the functional cataloging of microbial communities, which entails the deep metagenomic sequencing and de novo genomes assembly, thus providing better community structural and compositional as well as functional resolution of the microbial communities (Malla et al., 2018; Wei et al., 2018). Metatranscriptomics compliments metagenomics by allowing the identification of microbial genes that are expressed under different conditions. Meta proteomics and metabolomics help identify proteins and metabolites resulting from microbe-host co-metabolism, which can serve as reliable biomarkers given that they represent end products of metabolic interactions among the microbe and host. The combination of multi-omics technologies increases confidence in identified diagnostic and therapeutic biomarkers as well as provides testable hypotheses. To test emergent hypotheses generated as a result of these technologies and delineate mechanisms by which microbes influence the host, germ-free animal models provide a highly controllable experimental system with reduced complexity to study interactions between the host and its resident microbiota (Dubey et al., 2022a; Malla, Dubey, Raj et al., 2022).

2.8 CONCLUSION AND FUTURE PERSPECTIVES

Plants act as actual "filters" of soil microbial communities, identifying the productive and capable endophytes that not only provide a novel source of bioactive natural compounds but also provide an environmentally acceptable means of promoting plant growth. Now that a large number of microbial endophytes have been found, we can better understand the mechanisms underlying their symbiotic connection. Endophyte microbiome research is becoming more and more important as it relates to plant development, health, and production. Possibilities exist for utilizing microbial endophytes to improve host performance because these symbiotic communities are crucial for the control of stress in host plants. Even though the majority of research findings are basic, further study is required to unravel the complexities of the host-endophyte relationship to comprehend the interaction and dynamics of the microbial communities. As already mentioned, the endophytic microbial populations are essential to the functioning of the immunological and metabolic systems of plants. More thorough knowledge of these relationships will result from studying the host-endophyte interactions to the structural and functional characteristics of the microbial taxa. Furthermore, research into bacterial endophytes is still in its infancy. Nevertheless, omics-based methodologies have made it possible to study and understand these microbial symbionts. To benefit from this amazing relationship, it is imperative to study the physiology and biochemistry of these endophytic bacteria. Endophytes must be studied in the future to achieve the goals of environmental sustainability, agricultural productivity, and serving as a source of novel biomolecules for various aspects of human health directly related to and industrial sectors, while keeping their biological potential and role in plant growth and development in mind.

ACKNOWLEDGMENTS

AD received grants through the DST Inspire PhD Fellowship (IF160797) from the department of science and technology, New Delhi, India. DS and AK gratefully acknowledge DST-SERB for financial support obtained through a project grant of (CRG/2021/003696), New Delhi, Govt of India.

KEYWORDS

- **endophytes**
- **microbiome**
- **omics**
- **plant-endophyte interactions**
- **plant-microbiome sequencing**
- **resistance**
- **rhizosphere**
- **sustainable agriculture**

REFERENCES

Abraham, B. S., Caglayan, D., Carrillo, N. V., Chapman, M. C., Hagan, C. T., Hansen, S. T., Jeanty, R. O., et al., (2020). Shotgun metagenomic analysis of microbial communities from the Loxahatchee nature preserve in the Florida everglades. *Environmental Microbiomes, 15*(1). https://doi.org/10.1186/s40793-019-0352-4.

Agler, M. T., Ruhe, J., Kroll, S., Morhenn, C., Kim, S. T., Weigel, D., & Kemen, E. M., (2016). Microbial hub taxa link host and abiotic factors to plant microbiome variation. *PLoS Biology, 14*(1). https://doi.org/10.1371/journal.pbio.1002352.

Ahmad, T., Farooq, S., Mirza, D. N., Kumar, A., Mir, R. A., & Riyaz-Ul-Hassan, S., (2022). Insights into the endophytic bacterial microbiome of crocus sativus: Functional characterization leads to potential agents that enhance the plant growth, productivity, and key metabolite content. *Microbial Ecology, 83*, 669–688. https://doi.org/10.1007/s00248-021-01810-y.

Ali, S., Duan, J., Charles, T. C., & Glick, B. R., (2014). A bioinformatics approach to the determination of genes involved in endophytic behavior in *Burkholderia* spp. *Journal of Theoretical Biology, 343*, 193–198. https://doi.org/10.1016/j.jtbi.2013.10.007.

ALKahtani, M. D. F., Fouda, A., Attia, K. A., Al-Otaibi, F., Eid, A. M., El-Din, E. E., Hijri, M., et al., (2020). Isolation and characterization of plant growth promoting endophytic bacteria from desert plants and their application as bioinoculants for sustainable agriculture. *Agronomy, 10*. https://doi.org/10.3390/agronomy10091325.

Babalola, O. O., Fadiji, A. E., & Ayangbenro, A. S., (2020). Shotgun metagenomic data of root endophytic microbiome of maize (*Zea mays* L.). *Data in Brief, 31*. https://doi.org/10.1016/j.dib.2020.105893.

Balmer, A., Pastor, V., Gamir, J., Flors, V., & Mauch-Mani, B., (2015). The "prime-ome": Towards a holistic approach to priming. In: *Trends in Plant Science*. https://doi.org/10.1016/j.tplants.2015.04.002.

Berendsen, R. L., Van, V. M. C., Stringlis, I. A., Zamioudis, C., Tommassen, J., Pieterse, C. M. J., & Bakker, P. A. H. M., (2015). Unearthing the genomes of plant-beneficial *Pseudomonas* model strains WCS358, WCS374 and WCS417. *BMC Genomics*. https://doi.org/10.1186/s12864-015-1632-z.

Berendsen, R. L., Vismans, G., Yu, K., Song, Y., De Jonge, R., Burgman, W. P., Burmølle, M., et al., (2018). Disease-induced assemblage of a plant-beneficial bacterial consortium. *ISME Journal, 12*(6), 1496–1507. https://doi.org/10.1038/s41396-018-0093-1.

Bulgarelli, D., Garrido-Oter, R., Münch, P. C., Weiman, A., Dröge, J., Pan, Y., McHardy, A. C., & Schulze-Lefert, P., (2015). Structure and function of the bacterial root microbiota in wild and domesticated barley. *Cell Host and Microbe, 17*(3), 392–403. https://doi.org/10.1016/j.chom.2015.01.011.

Camacho, C., Coulouris, G., Avagyan, V., Ma, N., Papadopoulos, J., Bealer, K., & Madden, T. L., (2009). BLAST+: Architecture and applications. *BMC Bioinformatics, 10*. https://doi.org/10.1186/1471-2105-10-421.

Cameron, D. D., Neal, A. L., Van, W. S. C. M., & Ton, J., (2013). Mycorrhiza-induced resistance: More than the sum of its parts? In: *Trends in Plant Science* (Vol. 18, No. 10, pp. 539–545). https://doi.org/10.1016/j.tplants.2013.06.004.

Card, S., Johnson, L., Teasdale, S., & Caradus, J., (2016). Deciphering endophyte behavior: The link between endophyte biology and efficacious biological control agents. *FEMS Microbiology Ecology*. https://doi.org/10.1093/femsec/fiw114.

Carmody, R. N., Gerber, G. K., Luevano, J. M., Gatti, D. M., Somes, L., Svenson, K. L., & Turnbaugh, P. J., (2015). Diet dominates host genotype in shaping the murine gut microbiota. *Cell Host and Microbe*. https://doi.org/10.1016/j.chom.2014.11.010.

Chaparro, J. M., Badri, D. V., & Vivanco, J. M., (2014). Rhizosphere microbiome assemblage is affected by plant development. *ISME Journal*, *8*(4), 790–803. https://doi.org/10.1038/ismej.2013.196.

Copeland, J. K., Layeghifard, M., Wang, P. W., Yuan, L., & Guttman, D. S., (2015). Seasonal community succession of the phyllosphere microbiome. *Molecular Plant-Microbe Interactions*. https://doi.org/10.1094/mpmi-10-14-0331-fi.

Cord-Landwehr, S., Melcher, R. L. J., Kolkenbrock, S., & Moerschbacher, B. M., (2016). A chitin deacetylase from the endophytic fungus *Pestalotiopsis* sp. efficiently inactivates the elicitor activity of chitin oligomers in rice cells. *Scientific Reports*, *6*. https://doi.org/10.1038/srep38018.

Cui, J. L., Wang, Y. N., Jiao, J., Gong, Y., Wang, J. H., & Wang, M. L., (2017). Fungal endophyte-induced salidroside and tyrosol biosynthesis combined with signal cross-talk and the mechanism of enzyme gene expression in *Rhodiola crenulata*. *Scientific Reports*, *7*(1). https://doi.org/10.1038/s41598-017-12895-2.

Das, G., & Dhal, P. K., (2022). Salinity influences endophytic bacterial communities in rice roots from the Indian Sundarban area. *Current Microbiology*, *79*(8), 238. https://doi.org/10.1007/s00284-022-02936-z.

David, L. A., Maurice, C. F., Carmody, R. N., Gootenberg, D. B., Button, J. E., Wolfe, B. E., Ling, A. V., et al., (2014). Diet rapidly and reproducibly alters the human gut microbiome. *Nature*. https://doi.org/10.1038/nature12820.

De Medeiros, A. T., Aburjaile, F. F., Ferreira-Neto, J. R. C., Pandolfi, V., & Benko-Iseppon, A. M., (2021). The endophytome (plant-associated microbiome): Methodological approaches, biological aspects, and biotech applications. In: *World Journal of Microbiology and Biotechnology* (Vol. 37, No. 12). Springer Science and Business Media B.V. https://doi.org/10.1007/s11274-021-03168-2.

Di, Y.-ning., Kui, L., Singh, P., Liu, L. F., Xie, L. Y., He, L. L., & Li, F. S., (2022). Identification and characterization of *bacillus subtilis* B9: A diazotrophic plant growth-promoting endophytic bacterium isolated from sugarcane root. *Journal of Plant Growth Regulation*. https://doi.org/10.1007/s00344-022-10653-x.

Dubey, A., Kumar, A., Abd_Allah, E. F., Hashem, A., & Khan, M. L., (2019). Growing more with less: Breeding and developing drought resilient soybean to improve food security. *Ecological Indicators*, *105*, 425–437. https://doi.org/10.1016/j.ecolind.2018.03.003.

Dubey, A., Malla, M. A., & Kumar, A., (2022). Role of next-generation sequencing (NGS) in understanding the microbial diversity. In: *Molecular Genetics and Genomics Tools in Biodiversity Conservation* (pp. 307–328). Springer Nature Singapore. https://doi.org/10.1007/978-981-16-6005-4_16.

Dubey, A., Malla, M. A., & Kumar, A., (2022). Taxonomical and functional bacterial community profiling in disease-resistant and disease-susceptible soybean cultivars. *Brazilian Journal of Microbiology*. https://doi.org/10.1007/s42770-022-00746-w.

Dubey, A., Malla, M. A., Kumar, A., Dayanandan, S., & Khan, M. L., (2020). Plants endophytes: Unveiling hidden agenda for bioprospecting toward sustainable agriculture. In: *Critical Reviews in Biotechnology* (Vol. 40, No. 8, pp. 1210–1231). Taylor & Francis. https://doi.org/10.1080/07388551.2020.1808584.

Dubey, A., Saiyam, D., Kumar, A., Hashem, A., Abduallah, E. F., & Khan, M. L., (2021). Bacterial root endophytes: Characterization of their competence and plant growth promotion in soybean (*Glycine max* (L.) merr.) under drought stress. *International Journal of Environmental Research and Public Health, 18*(3), 1–20. https://doi.org/10.3390/ijerph18030931.

Edwards, J., Johnson, C., Santos-Medellín, C., Lurie, E., Podishetty, N. K., Bhatnagar, S., Eisen, J. A., et al., (2015). Structure, variation, and assembly of the root-associated microbiomes of rice. *Proceedings of the National Academy of Sciences of the United States of America, 112*(8), E911–E920. https://doi.org/10.1073/pnas.1414592112.

Enagbonma, B. J., Aremu, B. R., & Babalola, O. O., (2019). Profiling the functional diversity of termite mound soil bacteria as revealed by shotgun sequencing. *Genes, 10*(9). https://doi.org/10.3390/genes10090637.

Enquahone, S., Van, M. G., & Simachew, A., (2022). Plant growth-promoting characteristics of halotolerant endophytic bacteria isolated from *Sporobolus specatus* (Vahr) Kunth and *Cyperus laevigatus* L. of Ethiopian rift valley lakes. *Archives of Microbiology, 204*. https://doi.org/10.1007/s00203-022-03021-6.

Etminani, F., & Harighi, B., (2018). Isolation and identification of endophytic bacteria with plant growth promoting activity and biocontrol potential from wild pistachio trees. *Plant Pathology Journal, 34*(3), 208–217. https://doi.org/10.5423/PPJ.OA.07.2017.0158.

Fadiji, A. E., Ayangbenro, A. S., & Babalola, O. O., (2021). Shotgun metagenomics reveals the functional diversity of root-associated endophytic microbiomes in maize plant. *Current Plant Biology, 25*. https://doi.org/10.1016/j.cpb.2021.100195.

Fadiji, A. E., Ayansina, A. S., & Oluranti, B. O., (2021). *Unveiling the Putative Functional Genes Present in Root-Associated Endophytic Microbiome from Maize Plant Using the Shotgun Approach.* https://doi.org/10.1007/s13353-021-00611-w/Published.

Field, K. J., Pressel, S., Duckett, J. G., Rimington, W. R., & Bidartondo, M. I., (2015). Symbiotic options for the conquest of land. In: *Trends in Ecology and Evolution.* https://doi.org/10.1016/j.tree.2015.05.007.

Gilbert, J. A., & Dupont, C. L., (2011). Microbial metagenomics: Beyond the genome. *Annual Review of Marine Science, 3*, 347–371. https://doi.org/10.1146/annurev-marine-120709-142811.

Gilbert, S., Xu, J., Acosta, K., Poulev, A., Lebeis, S., & Lam, E., (2018). Bacterial production of indole related compounds reveals their role in association between duckweeds and endophytes. *Frontiers in Chemistry, 6*. https://doi.org/10.3389/fchem.2018.00265.

Gomes, M. L., Cordenonsi Da, F. G., Loss De, M. G., Felipe, V. D. O. L., Biso De, C. J., Rodrigues, K. F., & Margis, R., (2012). *Metatranscriptomic Analysis of Small RNAs Present in Soybean Deep Sequencing Libraries*. www.sbg.org.br.

Gutjahr, C., (2014). Phytohormone signaling in arbuscular mycorhiza development. In: *Current Opinion in Plant Biology* (Vol. 20, pp. 26–34). https://doi.org/10.1016/j.pbi.2014.04.003.

Hacquard, S., Garrido-Oter, R., González, A., Spaepen, S., Ackermann, G., Lebeis, S., McHardy, A. C., et al., (2015). Microbiota and host nutrition across plant and animal kingdoms. In: *Cell Host and Microbe* (Vol. 17, NO. 5, pp. 603–616). https://doi.org/10.1016/j.chom.2015.04.009.

Hadrich, D., (2018). Microbiome research is becoming the key to better understanding health and nutrition. In: *Frontiers in Genetics* (Vol. 9). https://doi.org/10.3389/fgene.2018.00212.

Hagagy, N., Hamedo, H., Elshafi, N., & Selim, S., (2021). Shotgun metagenomic sequencing of extremophilic community from Soda Lake, Ga'ar Lake, in Wadi Al-Natrun, Egypt.

The Egyptian Journal of Experimental Biology (Botany), *17*, 1. https://doi.org/10.5455/egyjebb.20210309063625.

Hardoim, P. R., Van, O. L. S., & Elsas, J. D. V., (2008). Properties of bacterial endophytes and their proposed role in plant growth. *Trends in Microbiology*, *16*(10), 463–471. https://doi.org/10.1016/j.tim.2008.07.008.

Hashem, A., Abd_allah, E. F., Alqarawi, A. A., Radhakrishnan, R., & Kumar, A., (2017). Plant defense approach of *Bacillus subtilis* (Bera 71) against *Macrophomina phaseolina* (tassi) goid in mung bean. *Journal of Plant Interactions*. https://doi.org/10.1080/17429145.2017.1373871.

Hunter, P., (2016). Plant microbiomes and sustainable agriculture. *EMBO Reports*, *17*(12), 1696–1699. https://doi.org/10.15252/embr.201643476.

Huson, D. H., & Weber, N., (2013). *Microbial Community Analysis Using Megan* (pp. 465–485). https://doi.org/10.1016/B978-0-12-407863-5.00021-6.

Ismail, M. A., Amin, M. A., Eid, A. M., Hassan, S. E. D., Mahgoub, H. A. M., Lashin, I., Abdelwahab, A. T., et al., (2021). Comparative study between exogenously applied plant growth hormones versus metabolites of microbial endophytes as plant growth-promoting for *Phaseolus Vulgaris* L. *Cells* 10. https://doi.org/10.3390/cells10051059.

Jia, H., Xi, Z., Ma, J., Li, Y., Hao, C., Lu, M., Zhang, Z. Z., & Deng, W. W., (2022). Endophytic bacteria from the leaves of two types of albino tea plants, indicating the plant growth promoting properties. *Plant Growth Regulation, 96*, 331–343. https://doi.org/10.1007/s10725-021-00779-5.

Jia, M., Chen, L., Xin, H. L., Zheng, C. J., Rahman, K., Han, T., & Qin, L. P., (2016). A friendly relationship between endophytic fungi and medicinal plants: A systematic review. In: *Frontiers in Microbiology* (Vol. 7). https://doi.org/cccc.

Jin, T., Shan, Z., Zhou, S., Yang, Q., Gai, J., & Li, Y., (2022). GmDNAJC7 from soybean is involved in plant tolerance to alkaline-salt, salt, and drought stresses. *Agronomy*, *12*(6), 1419. https://doi.org/10.3390/agronomy12061419.

Johnson, M., Zaretskaya, I., Raytselis, Y., Merezhuk, Y., McGinnis, S., & Madden, T. L., (2008). NCBI BLAST: A better web interface. *Nucleic Acids Research*, *36*(Web Server Issue). https://doi.org/10.1093/nar/gkn201.

Kane, K. H., (2011). Effects of endophyte infection on drought stress tolerance of *Lolium perenne* accessions from the Mediterranean region. *Environmental and Experimental Botany*. https://doi.org/10.1016/j.envexpbot.2011.01.002.

Kaul, S., Sharma, T., & Dhar, M. K., (2016). "Omics" tools for better understanding the plant–endophyte interactions. *Frontiers in Plant Science*, *7*. https://doi.org/10.3389/fpls.2016.00955.

Khanna, A., Raj, K., Kumar, P., & Wati, L., (2022). Antagonistic and growth-promoting potential of multifarious bacterial endophytes against *Fusarium* wilt of chickpea. *Egyptian Journal of Biological Pest Control, 32*. https://doi.org/10.1186/s41938-022-00516-8.

Khare, E., Mishra, J., & Arora, N. K., (2018). Multifaceted interactions between endophytes and plant: Developments and prospects. In: *Frontiers in Microbiology* (Vol. 9). https://doi.org/10.3389/fmicb.2018.02732.

Kong, Z., Mohamad, O. A., Deng, Z., Liu, X., Glick, B. R., & Wei, G., (2015). Rhizobial symbiosis effect on the growth, metal uptake, and antioxidant responses of *Medicago lupulina* under copper stress. *Environmental Science and Pollution Research*, *22*(16), 12479–12489. https://doi.org/10.1007/s11356-015-4530-7.

Kostic, A. D., Howitt, M. R., & Garrett, W. S., (2013). Exploring host-microbiota interactions in animal models and humans. *Genes and Development*. https://doi.org/10.1101/gad.212522.112.

Kumar, A., & Dubey, A., (2020). Rhizosphere microbiome: Engineering bacterial competitiveness for enhancing crop production. In: *Journal of Advanced Research* (Vol. 24, pp. 337–352). Elsevier B.V. https://doi.org/10.1016/j.jare.2020.04.014.

Kumar, S., Krishnani, K. K., Bhushan, B., & Brahmane, M. P., (2015). Metagenomics: Retrospect and prospects in high throughput age. *Biotechnology Research International, 2015*, 1–13. https://doi.org/10.1155/2015/121735.

Lastochkina, O., Pusenkova, L., Garshina, D., Kasnak, C., Palamutoglu, R., Shpirnaya, I., Mardanshin, I., & Maksimov, I., (2022). Improving the biocontrol potential of endophytic bacteria *Bacillus subtilis* with salicylic acid against phytophthora infestans-caused postharvest potato tuber late blight and impact on stored tubers quality. *Horticulturae*, *8*(2). https://doi.org/10.3390/horticulturae8020117.

Lata, R., Chowdhury, S., Gond, S. K., & White, J. F., (2018). Induction of abiotic stress tolerance in plants by endophytic microbes. In: *Letters in Applied Microbiology* (Vol. 66, No. 4, pp. 268–276). https://doi.org/10.1111/lam.12855.

Li, H. Y., Wei, D. Q., Shen, M., & Zhou, Z. P., (2012). Endophytes and their role in phytoremediation. In: *Fungal Diversity*. https://doi.org/10.1007/s13225-012-0165-x.

Lugtenberg, B. J. J., Caradus, J. R., & Johnson, L. J., (2016). Fungal endophytes for sustainable crop production. In: *FEMS Microbiology Ecology* (Vol. 92, No. 12). https://doi.org/10.1093/femsec-fiw194.

Mahgoub, H. A. M., Fouda, A., Eid, A. M., Ewais, E. E. D., & Hassan, S. E. D., (2021). Biotechnological application of plant growth-promoting endophytic bacteria isolated from halophytic plants to ameliorate salinity tolerance of *Vicia faba* L. *Plant Biotechnology Reports, 15,* 819–843. https://doi.org/10.1007/s11816-021-00716-y.

Malla, M. A., Dubey, A., Kumar, A., & Yadav, S., (2022). Metagenomic analysis displays the potential predictive biodegradation pathways of the persistent pesticides in agricultural soil with a long record of pesticide usage. *Microbiological Research, 261*, 127081. https://doi.org/10.1016/j.micres.2022.127081.

Malla, M. A., Dubey, A., Kumar, A., Yadav, S., Hashem, A., & Abd_Allah, E. F., (2019). *Exploring the Human Microbiome: The Potential Future Role of Next-Generation Sequencing in Disease Diagnosis and Treatment*, *9*(2868). https://doi.org/10.3389/fimmu.2018.02868.

Malla, M. A., Dubey, A., Raj, A., Kumar, A., Upadhyay, N., & Yadav, S., (2022). Emerging frontiers in microbe-mediated pesticide remediation: Unveiling role of omics and in silico approaches in engineered environment. *Environmental Pollution (Barking, Essex: 1987), 299*, 118851. https://doi.org/10.1016/j.envpol.2022.118851.

Malla, M. A., Dubey, A., Yadav, S., Kumar, A., Hashem, A., & Abd-Allah, E. F., (2019). Exploring the human microbiome: The potential future role of next-generation sequencing in disease diagnosis and treatment. *Frontiers in Immunology*, *9*(19), 1–23. https://doi.org/10.3389/fimmu.2018.02868.

Malla, M. A., Dubey, A., Yadav, S., Kumar, A., Hashem, A., & Abd, A. E. F., (2018). Understanding and designing the strategies for the microbe-mediated remediation of environmental contaminants using omics approaches. *Frontiers in Microbiology*, *9*. https://doi.org/10.3389/fmicb.2018.01132.

Marchesi, J. R., & Ravel, J., (2015). The vocabulary of microbiome research: A proposal. *Microbiome, 3*(1). https://doi.org/10.1186/s40168-015-0094-5.

Martínez-Hidalgo, P., Maymon, M., Pule-Meulenberg, F., & Hirsch, A. M., (2018). Engineering root microbiomes for healthier crops and soils using beneficial, environmentally safe bacteria. *Canadian Journal of Microbiology*. https://doi.org/10.1139/cjm-2018-0315.

Mei, C., Chretien, R. L., Amaradasa, B. S., He, Y., Turner, A., & Lowman, S., (2021). Characterization of phosphate solubilizing bacterial endophytes and plant growth promotion *in vitro* and in greenhouse. *Microorganisms, 9*. https://doi.org/10.3390/microorganisms9091935.

Melcher, U., Verma, R., & Schneider, W. L., (2014). Metagenomic search strategies for interactions among plants and multiple microbes. In: *Frontiers in Plant Science* (Vol. 5). Frontiers Research Foundation. https://doi.org/10.3389/fpls.2014.00268.

Mendes, L. W., Mendes, R., Raaijmakers, J. M., & Tsai, S. M., (2018). Breeding for soil-borne pathogen resistance impacts active rhizosphere microbiome of common bean. *ISME Journal*. https://doi.org/10.1038/s41396-018-0234-6.

Mendes, R., Kruijt, M., De Bruijn, I., Dekkers, E., Van, D. V. M., Schneider, J. H. M., Piceno, Y. M., et al., (2011). Deciphering the rhizosphere microbiome for disease-suppressive bacteria. *Science, 332*(6033), 1097–1100. https://doi.org/10.1126/science.1203980.

Mesa, V., Navazas, A., González-Gil, R., González, A., Weyens, N., Lauga, B., Gallego, J. L. R., et al., (2017). Use of endophytic and rhizosphere bacteria to improve phytoremediation of arsenic-contaminated industrial soils by autochthonous *Betula celtiberica. Applied and Environmental Microbiology*. https://doi.org/10.1128/AEM.03411-16.

Milone, D. H., Stegmayer, G. S., Kamenetzky, L., López, M., Lee, J. M., Giovannoni, J. J., & Carrari, F. (2010). omeSOM: a software for clustering and visualization of transcriptional and metabolite data mined from interspecific crosses of crop plants. *BMC Bioinformatics, 11*(1), 1–10. https://doi.org/10.1186/1471-2105-11-438.

Newman, M. A., Sundelin, T., Nielsen, J. T., & Erbs, G., (2013). MAMP (microbe-associated molecular pattern) triggered immunity in plants. *Frontiers in Plant Science, 4*. https://doi.org/10.3389/fpls.2013.00139.

Ofek-Lalzar, M., Sela, N., Goldman-Voronov, M., Green, S. J., Hadar, Y., & Minz, D., (2014). Niche and host-associated functional signatures of the root surface microbiome. *Nature Communications, 5*. https://doi.org/10.1038/ncomms5950.

Pandey, V., Ansari, M. W., Tula, S., Yadav, S., Sahoo, R. K., Shukla, N., Bains, G., et al., (2016). Dose-dependent response of *Trichoderma harzianum* in improving drought tolerance in rice genotypes. *Planta, 243*(5), 1251–1264. https://doi.org/10.1007/s00425-016-2482-x.

Paolinelli, M., Escoriaza, G., Cesari, C., Garcia-Lampasona, S., & Hernandez-Martinez, R., (2022). Characterization of grapevine wood microbiome through a metatranscriptomic approach. *Microbial Ecology, 83*(3), 658–668. https://doi.org/10.1007/s00248-021-01801-z.

Pappas, M. L., Liapoura, M., Papantoniou, D., Avramidou, M., Kavroulakis, N., Weinhold, A., Broufas, G. D., & Papadopoulou, K. K., (2018). The beneficial endophytic fungus *fusarium* solani strain k alters tomato responses against spider mites to the benefit of the plant. *Frontiers in Plant Science, 9*. https://doi.org/10.3389/fpls.2018.01603.

Patle, P., Phule, K. V. M., Navnage, I. N., Krishi, V. G., Ramteke, I. P., Navnage, N., & Ramteke, P., (2018). Endophytes in plant system: Roles in growth promotion, mechanism, and their potentiality in achieving agriculture sustainability. *International Journal of Chemical Studies*.

Peiffer, J., Spor, A., Koren, O., Jin, Z., Tringe, S. G., Dangl, J. L., Buckler, E. S., & Ley, R. E., (2013). Diversity and heritability of the maize rhizosphere microbiome under field conditions. *Proceedings of the National Academy of Sciences*. https://doi.org/10.1073/pnas.1302837110.

Pieterse, C. M. J., Zamioudis, C., Berendsen, R. L., Weller, D. M., Van, W. S. C. M., & Bakker, P. A. H. M., (2014). Induced systemic resistance by beneficial microbes. *Annual Review of Phytopathology, 52*(1), 347–375. https://doi.org/10.1146/annurev-phyto-082712-102340.

Porras-Alfaro, A., & Bayman, P., (2011). Hidden fungi, emergent properties: Endophytes and microbiomes. *Annual Review of Phytopathology*. https://doi.org/10.1146/annurev-phyto-080508-081831.

Pozo, M. J., & Azcón-Aguilar, C., (2007). Unraveling mycorrhiza-induced resistance. In: *Current Opinion in Plant Biology*. https://doi.org/10.1016/j.pbi.2007.05.004.

Pozo, M. J., López-Ráez, J. A., Azcón-Aguilar, C., & García-Garrido, J. M., (2015). Phytohormones as integrators of environmental signals in the regulation of mycorrhizal symbioses. In: *New Phytologist* (Vol. 205, No. 4, pp. 1431–1436). https://doi.org/10.1111/nph.13252.

Raghavendra, A. K. H., & Newcombe, G., (2013). The contribution of foliar endophytes to quantitative resistance to *Melampsora* rust. *New Phytologist*. https://doi.org/10.1111/nph.12066.

Ramírez-puebla, S. T., Servín-Garcidueñas, L. E., Jiménez-Marín, B., Bolaños, L. M., Rosenblueth, M., Martínez, J., Rogel, M. A., et al., (2013). Gut and root microbiota commonalities. In: *Applied and Environmental Microbiology* (Vol. 79, No. 1, pp. 2–9). https://doi.org/10.1128/AEM.02553-12.

Sahu, P. K., Gupta, A., Lavanya, G., Bakade, R., & Singh, D. P., (2017). Bacterial endophytes: Potential candidates for plant growth promotion. In: *Plant-Microbe Interactions in Agro-Ecological Perspectives* (Vol. 1, pp. 633–657). https://doi.org/10.1007/978-981-10-5813-4_31.

Salam, N., Khieu, T. N., Liu, M. J., Vu, T. T., Chu-Ky, S., Quach, N. T., Phi, Q. T., et al., (2017). Endophytic actinobacteria associated with *Dracaena cochinchinensis* Lour.: Isolation, diversity, and their cytotoxic activities. *BioMed Research International*. https://doi.org/10.1155/2017/1308563.

Saminathan, T., García, M., Ghimire, B., Lopez, C., Bodunrin, A., Nimmakayala, P., Abburi, V. L., et al., (2018). Metagenomic and metatranscriptomic analyses of diverse watermelon cultivars reveal the role of fruit associated microbiome in carbohydrate metabolism and ripening of mature fruits. *Frontiers in Plant Science, 9*. https://doi.org/10.3389/fpls.2018.00004.

Sánchez-Vallet, A., Mesters, J. R., & Thomma, B. P. H. J., (2015). The battle for chitin recognition in plant-microbe interactions. In: *FEMS Microbiology Reviews* (Vol. 39, No. 2, pp. 171–183). https://doi.org/10.1093/femsre/fuu003.

Schardl, C. L., Young, C. A., Pan, J., Florea, S., Takach, J. E., Panaccione, D. G., Farman, M. L., et al., (2013). Currencies of mutualisms: Sources of alkaloid genes in vertically transmitted epichloae. *Toxins*. https://doi.org/10.3390/toxins5061064.

Schlaeppi, K., Ver, L. V. T. E., Schulze-Lefert, P., Oter, R. G., & Dombrowski, N., (2013). Quantitative divergence of the bacterial root microbiota in *Arabidopsis thaliana* relatives. *Proceedings of the National Academy of Sciences*. https://doi.org/10.1073/pnas.1321597111.

Schreiter, S., Ding, G. C., Heuer, H., Neumann, G., Sandmann, M., Grosch, R., Kropf, S., & Smalla, K., (2014). Effect of the soil type on the microbiome in the rhizosphere of field-grown lettuce. *Frontiers in Microbiology, 5*. https://doi.org/10.3389/fmicb.2014.00144.

Sessitsch, A., Hardoim, P., Döring, J., Weilharter, A., Krause, A., Woyke, T., Mitter, B., et al., (2012). Functional characteristics of an endophyte community colonizing rice roots as revealed by metagenomic analysis. *Molecular Plant-Microbe Interactions MPMI, 25*(1), 28–36. https://doi.org/10.1094/MPMI.

Shakya, M., Gottel, N., Castro, H., Yang, Z. K., Gunter, L., Labbé, J., Muchero, W., et al., (2013). A Multifactor analysis of fungal and bacterial community structure in the root microbiome of mature *Populus deltoides* Trees. *PLoS One, 8*(10). https://doi.org/10.1371/journal.pone.0076382.

Sheoran, N., Valiya, N. A., Munjal, V., Kundu, A., Subaharan, K., Venugopal, V., Rajamma, S., et al., (2015). Genetic analysis of plant endophytic *Pseudomonas putida* BP25 and chemo-profiling of its antimicrobial volatile organic compounds. *Microbiological Research, 173*, 66–78. https://doi.org/10.1016/j.micres.2015.02.001.

Shi, S., Nuccio, E., Herman, D. J., Rijkers, R., Estera, K., Li, J., Da Rocha, U. N., et al., (2015). Successional trajectories of rhizosphere bacterial communities over consecutive seasons. *MBio*. https://doi.org/10.1128/mBio.00746-15.

Shurigin, V., Alikulov, B., Davranov, K., & Ismailov, Z., (2022). Bacterial endophytes from halophyte black saxaul (*Haloxylon aphyllum* Minkw.) and their plant growth-promoting properties. *Journal of Applied Biology and Biotechnology, 10*, 45–53. https://doi.org/10.7324/JABB.2021.100106.

Singh, R., Pandey, K. D., Singh, M., Singh, S. K., Hashem, A., Al-Arjani, A. B. F., Abd_Allah, E. F., et al., (2022). Isolation and characterization of endophytes bacterial strains of *Momordica charantia* L. and their possible approach in stress management. *Microorganisms, 10*. https://doi.org/10.3390/microorganisms10020290.

Sprent, J. I., (2008). 60Ma of legume nodulation. What's new? What's changing? *Journal of Experimental Botany, 59*(5), 1081–1084. https://doi.org/10.1093/jxb/erm286.

Staerck, C., Gastebois, A., Vandeputte, P., Calenda, A., Larcher, G., Gillmann, L., Papon, N., et al., (2017). Microbial antioxidant defense enzymes. In: *Microbial Pathogenesis* (Vol. 110, pp. 56–65). https://doi.org/10.1016/j.micpath.2017.06.015.

Sugiyama, A., Ueda, Y., Zushi, T., Takase, H., & Yazaki, K., (2014). Changes in the bacterial community of soybean rhizospheres during growth in the field. *PLoS One, 9*(6), 1–9. https://doi.org/10.1371/journal.pone.0100709.

Tao, L., Qiuhong, L., Fuqiang, Y., Shuhui, Z., Suohui, T., & Linyuan, F., (2022). Plant growth-promoting activities of bacterial endophytes isolated from the medicinal plant *Pairs polyphylla* var. yunnanensis. *World Journal of Microbiology and Biotechnology, 38*. https://doi.org/10.1007/s11274-021-03194-0.

Trdá, L., Boutrot, F., Claverie, J., Brulé, D., Dorey, S., & Poinssot, B., (2015). Perception of pathogenic or beneficial bacteria and their evasion of host immunity: Pattern recognition receptors in the frontline. *Frontiers in Plant Science*. https://doi.org/10.3389/fpls.2015.00219.

Tringe, S. G., Von, M. C., Kobayashi, A., Salamov, A. A., Chen, K., Chang, H. W., Podar, M., et al., (2005). Comparative metagenomics of microbial communities. *Science, 308*(5721), 554–557.

Tsalgatidou, P. C., Thomloudi, E. E., Baira, E., Papadimitriou, K., Skagia, A., Venieraki, A., & Katinakis, P., (2022). Integrated genomic and metabolomic analysis illuminates key secreted metabolites produced by the novel endophyte *Bacillus halotolerans* Cal.l.30 involved in diverse biological control activities. *Microorganisms, 10*(2). https://doi.org/10.3390/microorganisms10020399.

Turner, T. R., James, E. K., Poole, P. S., Gilbert, J., Meyer, F., Jansson, J., Gordon, J., et al., (2013). The plant microbiome. *Genome Biology, 14*(6), 209. https://doi.org/10.1186/gb-2013-14-6-209.

Van, D. E. S., Van, H. M., Pozo, M. J., Czechowski, T., Udvardi, M. K., Pieterse, C. M. J., & Ton, J., (2009). Priming of plant innate immunity by rhizobacteria and β-aminobutyric acid: Differences and similarities in regulation. *New Phytologist*. https://doi.org/10.1111/j.1469-8137.2009.02851.x.

Van, D. H. M. G. A., & Hartmann, M., (2016). Networking in the plant microbiome. In: *PLoS Biology* (Vol. 14, No. 2). https://doi.org/10.1371/journal.pbio.1002378.

Vandenkoornhuyse, P., Quaiser, A., Duhamel, M., Le Van, A., & Dufresne, A., (2015). The importance of the microbiome of the plant holobiont. In: *New Phytologist* (Vol. 206, No. 4, pp. 1196–1206). https://doi.org/10.1111/nph.13312.

Verma, S., Kumar, M., Kumar, A., Das, S., Chakdar, H., Varma, A., & Saxena, A. K., (2022). Diversity of bacterial endophytes of maize (*Zea mays*) and their functional potential for micronutrient biofortification. *Current Microbiology, 79*. https://doi.org/10.1007/s00284-021-02702-7.

Waghunde, R. R., Shelake, R. M., Shinde, M. S., & Hayashi, H., (2017). *Endophyte Microbes: A Weapon for Plant Health Management*. https://doi.org/10.1007/978-981-10-6241-4_16.

Wani, S. H., Kumar, V., Shriram, V., & Sah, S. K., (2016). Phytohormones and their metabolic engineering for abiotic stress tolerance in crop plants. In: *Crop Journal* (Vol. 4, No. 3, pp. 162–176). https://doi.org/10.1016/j.cj.2016.01.010.

Waqas, M., Khan, A. L., Kamran, M., Hamayun, M., Kang, S. M., Kim, Y. H., & Lee, I. J., (2012). Endophytic fungi produce gibberellins and indoleacetic acid and promotes host-plant growth during stress. *Molecules, 17*(9), 10754–10773. https://doi.org/10.3390/molecules170910754.

Waqas, M., Khana, A. L., Hamayuna, M., Shahzad, R., Kang, S. M., Kim, J. G., & Lee, I. J., (2015). Endophytic fungi promote plant growth and mitigate the adverse effects of stem rot: An example of *Penicillium citrinum* and *Aspergillus terreus*. *Journal of Plant Interactions, 10*(1), 280–287. https://doi.org/10.1080/17429145.2015.1079743.

Wei, Y. J., Wu, Y., Yan, Y. Z., Zou, W., Xue, J., Ma, W. R., Wang, W., et al., (2018). High-throughput sequencing of microbial community diversity in soil, grapes, leaves, grape juice, and wine of grapevine from China. *PLoS One, 13*(3). https://doi.org/10.1371/journal.pone.0193097.

Xu, H., Gao, J., Portieles, R., Du, L., Gao, X., & Borras-Hidalgo, O., (2022). Endophytic bacterium *Bacillus aryabhattai* induces novel transcriptomic changes to stimulate plant growth. *Plos One, 17*, e0272500. https://doi.org/10.1371/journal.pone.0272500.

Xu, R., Rajeev, S., & Salvador, L. C. M., (2022.). Selection of software and database for metagenomics sequence analysis impacts the outcome of microbial profiling and pathogen detection. *Journal of Applied Microbiology*. https://doi.org/10.1101/2022.04.27.489679.

Xu, Y., Liu, F., Zhu, S., & Li, X., (2018). The maize NBS-LRR Gene ZmNBS25 enhances disease resistance in rice and *Arabidopsis*. *Frontiers in Plant Science, 9*. https://doi.org/10.3389/fpls.2018.01033.

Yu, S., Yang, F., Zou, Y., Yang, Y., Li, T., Chen, S., Wang, Y., et al., (2022). Overexpressing PpBURP2 in rice increases plant defense to abiotic stress and bacterial leaf blight. *Frontiers in Plant Science, 13*. https://doi.org/10.3389/fpls.2022.812279.

Zarraonaindia, I., Owens, S. M., Weisenhorn, P., West, K., Hampton-Marcell, J., Lax, S., Bokulich, N. A., et al., (2015). The soil microbiome influences grapevine-associated microbiota. *MBio*. https://doi.org/10.1128/mBio.02527-14.

CHAPTER 3

Phyto-Microbiomes: Role in Sustainable Agriculture Practices Under Abiotic Stress

ANAMIKA DUBEY,[1] AMAN RAJ,[1] MUNEER AHMAD MALLA,[1] MOHAMMAD LATIF KHAN,[1] and ASHWANI KUMAR[1,2]

[1]*Metagenomics and Secretomics Research Laboratory, Department of Botany, Dr. Harisingh Gour Vishwavidyalaya (A Central University), Sagar, Madhya Pradesh, India*

[2]*Metagenomics and Secretomics Research Laboratory, Department of Botany, University of Allahabad (A Central University), Prayagraj, Uttar Pradesh, India*

ABSTRACT

Providing food security while remaining ecologically sustainable is a worldwide concern. To achieve this, agricultural productivity must increase by 70% under more difficult climatic circumstances while preserving environmental quality (e.g., reduced use of agrochemicals). To address this, the majority of governments and intergovernmental organizations have emphasized the need for alternative ways that leverage natural resources. The use of beneficial phytomicrobiomes (microbes that are intimately connected with plant tissues) is seen as one of the feasible answers to the dual concerns of food security and environmental sustainability. A variety of vital microorganisms are present in numerous areas of the plant, including the leaf, shoot, root, flower, and seed play important roles in plant health, growth, and productivity, as well as potentially contributing directly to enhancing food quality and quantity. The phytomicrobiome can also boost production by improving resource use efficiency and resistance to biotic and abiotic stressors. In this

Advances in Plant Microbiome Research for Climate Resilient Agriculture: Toward Sustainable Farming. Ashwani Kumar (Ed.)

chapter, we look at the role of phytomicrobiome in plant health and how functional microbiome features might be used to boost agricultural output in environmentally beneficial ways. However, important technical and translational obstacles remain, such as the variability of microbial product performance in field circumstances and a lack of techniques to modify the microbiome *in situ.* We offer approaches that necessitate a system-based approach to achieve the phytomicrobiome's potential in contributing to food security. We believe that if these technological and translational obstacles are addressed systematically, phytomicrobiome can greatly contribute to the long-term growth in agricultural production and food security.

3.1 INTRODUCTION

The global population is experiencing exponential growth, and it is anticipated to achieve 9.7 billion by 2050. Food production must increase by 70% to satisfy this expanding population (FAO, 2009). This rise in food demand must be fulfilled using current cultivable land and under more difficult climate conditions without further deterioration of ecosystems (Singh et al., 2020). Conversely, arable land is needed to help fulfill other social issues such as climate change mitigation and biodiversity protection (FAO, 2020). Furthermore, traditional agricultural systems that depend on elevated use of chemical fertilizer and pesticides have been criticized for their negative impact on environmental sustainability and food quality (Busby et al., 2016), including high chemical residues in food that harm human health and negative pesticide impacts on non-target organisms (e.g., bees, fish, and birds). This is exacerbated by the structural reduction in farm production in many regions of the world as a result of chemical effectiveness and plant breeding programs to boost yield. This indicates that increasing chemical inputs will not result in a corresponding rise in farm production (Trivedi et al., 2020). Most governmental and non-governmental organizations (e.g., FAO) have emphasized the need of developing and implementing nature-based solutions to boost agricultural output in an ecologically friendly manner. For example, it has been proposed that combining efforts to improve soil biological health and plant breeding programs can produce a 50% improvement in agricultural output without raising inputs (Dubey et al., 2019). Microbes are pervasive in the biosphere, contributing to the survival of the Earth's life system. As a result of millions of years of co-evolution between hosts and their associated bacteria, host-associated microorganisms play a key role in host fitness and health (Berendsen et al., 2012; Ruzzi et al., 2016; Yadav et al., 2020). Many studies conducted in recent years have highlighted the roles of plant growth promoting microbes in improving the growth and performance of different crop plants (Prudent et al., 2015; Dubey et al., 2021, 2022), wheat (Gontia-Mishra et al., 2016; Rascovan et al., 2016a), maize (Vardharajula et al., 2011; Sandhya et al., 2017; Oliveira et al., 2018), barley (Ghaffari et al., 2019), etc. This has resulted in multiple concepts that offer diverse techniques for harnessing phytomicrobiome to boost resource efficiency and availability, as well as protection against different pests and diseases for sustainable agriculture production (Maxton et al., 2018; Trivedi et al., 2020). The utilization of microbes and microbiomes has attracted the attention of both researchers and industries, and the agriculture microbiome is presently one of the fastest-growing sectors in the world, rising at a pace of 17% per year. This expansion in the agricultural microbiome market is attributed to an increase in public demand for less chemical-residue food, governmental requirements (e.g., the EU Green Deal demands 50% reduced use of synthetic insecticides by 2030), and environmental sustainability. A large number of microbes colonize the rhizosphere (0.54 mm) of plant roots (Kudoyarova et al., 2014; Fasusi et al., 2021).

The rhizosphere microbiome, which includes archaea, bacteria, viruses, fungi, and many other types of microorganisms, and their interactions, are essential for the absorption of plant nutrients as well as defense against a wide range of pests and diseases (Lakshmanan et al., 2014). There are several other advantages of microbes on plants, such as growth stimulation (e.g., phytohormones) or protection against abiotic (e.g., drought tolerance) and biotic (e.g., pathogen invasion) challenges. Plants secrete substantial quantities of exudates (consisting of up to 40% of photosynthesized carbons), which contain amino acids and sugar and serve as a source of nutrition for microbes in the rhizosphere. Similarly, plants harbor a diverse range of microbial phylotypes within plant tissues (both extracellularly and intracellularly), which are referred to as endophytes. Endophytes have more close connections with plants than free microorganisms in the soil (Dubey et al., 2020b, 2021). In addition to the use of microbial inoculants, *in situ* modification of plant microbiome has the potential to minimize abiotic and biotic stress in crops and therefore boost agricultural output (Miransari, 2016). Such techniques have the potential to produce a climate-resilient, sustainable agriculture practice that boosts agricultural production while also improving environmental sustainability (Yergeau et al., 2015; Malla et al., 2018, 2019; Krishna et al., 2019). However, crucial information gaps remain in both fundamental applied and translational research, limiting the development of effective interventions.

The functional features of the phytomicrobiome are highlighted in this chapter, with a focus on cultivable bacteria that may be exploited to promote sustainable farming. Like (i) the colonization process of phytomicrobiome;

(ii) emphasize the mechanism and functional aspects of phytomicrobiome; and (iii) present the evolving approach to phytomicrobiome problems and future outlooks.

3.2 PHYTOMICROBIOME AND HOST FUNCTIONS

Most of the above-ground (leaf, stem, seed, and flower) and below-ground (root) plant parts are highly populated by a different group of microbes that work together and influence plant health, just like the human body. The phytomicrobiome associated with various plant parts is distinctive. At the community level, the rhizomicrobiome (root-associated microbiome), endo-microbiome (microbes colonizing inner plant tissue) and phyllomicrobiome (leaf-associated microbes) all are distinct. Plant species, developmental phases of the same crop, and the environmental conditions all shape phyto-microbiomes (Quiza et al., 2015; Chouhan et al., 2021b).

Previous studies indicated that the root microbiome is dominated by *Actinobacteria, Acidobacteria, Proteobacteria, Verrucomicrobia, Bacteroidetes, Firmicutes, Gemmatimonadetes*, and *Planctomycetes* (Zarraonaindia et al., 2015; Anwar et al., 2016; Bhattacharyya et al., 2018; Compant et al., 2019; Lee et al., 2021). Aboveground plant parts (leaf, shoot, seed, and flower) seem to be dominated by *Proteobacteria, Actinobacteria, Firmicutes*, and *Bacteroidetes* (Zarraonaindia et al., 2015; Steven et al., 2018; Compant et al., 2019). We know very little about the ecological mechanisms that support the formation of microbiomes in diverse plant sections. Emerging findings revealed that several phytomicrobiome members are present in the majority of samples of a certain plant species regardless of climatic, geographical, or managerial factors, which are referred to as the core microbiome (Chouhan et al., 2021b). Members of core microbiomes are thought to be transmitted or explicitly recruited by a plant species, implying that these microorganisms co-evolved about plant species. It is suggested that some core microbiome members are inherited (through seed) and then influence other members' genotypes of plants, ambient circumstances, and management approaches (Chouhan et al., 2021a). The core microbiome enabled for identifying plant-associated microorganisms, which may be focused on further research and development for the endophytic microbe, genetically modified plants to improve nutritional value quality. Metagenomic and proteomic analysis of endophytic microorganisms, both uncultured and cultivated, have a functional role in plants can give specific details regarding the plant-microbe interaction as well as its functioning phenomena under varied environmental conditions.

3.3 ROOT-ASSOCIATED MICROBIOME

The root-associated microbiomes have garnered the greatest research to better understand their functions in delivering nutrients, immunity, and sustainability in harsh environments (Fitzpatrick et al., 2018). It is hypothesized that root microbiomes have a vital and similar function in plant health and fitness to gut microbiomes, including nutrition availability, defense against invading pests and diseases, and resistance to stressors (Berendsen et al., 2012; Kumar & Dubey, 2020; Dubey et al., 2021). It is suggested that the plant selects root microbiomes by structural alterations and root exudation of sugar, amino acids, and organic acids. Microbes, on the other hand, choose plants based on their capacity to colonize and compete for resources, as well as their ability to give resistance to abiotic and biotic challenges (Trivedi et al., 2020).

Root-associated microbiomes are more diversified than interactions with soil, which serves as a microbial seed bank for plants. Actinobacteria, Bacteroidetes, and Proteobacteria are common microbial components in plant roots, but Gemmatimonadetes, Acidobacteria, Archaea, and other microbial phylotypes are lacking. When compared to bulk soils, the sugarcane microbiome was shown to be dominated by Bacteroidetes, Proteobacteria, Actinobacteria, and reduced in Gemmatimonadetes and Acidobacteria. In root locations that differed considerably from bulk soils, Ascomycota was found to be the most prevalent fungal phyla, followed by Zygomycota, Glomeromycota, Basidiomycota, Chytridiomycota, and Rozellomycota. Some root microbiome investigations validate the previously reported enrichment and depletion of particular microbial species in plant roots, including Arabidopsis (Engelbrektson et al., 2012; Pétriacq et al., 2017), Tomato (Kwak et al., 2018), Maize (Peiffer et al., 2013), lettuce (Schreiter et al., 2014), barley (Bulgarelli et al., 2015), sugarcane (Yeoh et al., 2016), rice (Edwards et al., 2015), wheat (Rascovan et al., 2016b; Chen et al., 2019; Wolinska et al., 2020), sorghum (Kumar et al., 2021) Canola, lentil, and pea (Cordero et al., 2020).

3.4 PHYLLOSPHERE MICROBIOME

The phyllospheric microbes can sustain high temperatures and UV radiations, so they come under the class of extremophiles (Busby et al., 2016). A varied microorganism colonizes the plant phyllosphere, which has been observed to be unique to each plant species, rhizosphere, and endosphere (Dubey et al., 2020b, 2021). Phyllosphere consists of the surfaces of the stem (caulosphere), leaf, fruit (carposphere), flower, anther (anthosphere), as well as their endosphere (Figure 3.1). Flowers have close ties to pollinators, and nectar microbiomes have been shown to impact health and behavior of pollinator, resulting in increased scientific interest (Liu et al., 2019). Abiotic changes in the phyllosphere microbiome, such as climate change, increasing humidity, radiation levels, food availability, and plant genotype, might alter the cultivar-dependent selection of plant microbiomes both directly and indirectly (Timmusk et al., 2012; Dhankher & Foyer, 2018).

The existence and proliferation of the phyllosphere leaf microorganisms is largely determined by leaf exudates like sucrose, fructose, glucose, and amino acids. Dominant phyllospheric microbes include *Azotobacter, Alcaligenes, Arthrobacter, Agrobacterium, Achromobacter, Bacillus, Brevundimonas, Micromonospora, Methylobacterium, Micrococcus, Corynebacterium, Enterobacter, Haemophilus, Lysinibacillus, Pantoea, Pseudomonas*, *Psychrobacter Paenibacillus, Streptomyces*, and *Stenotrophomonas* (Mahapatra et al., 2020).

3.4.1 LEAF MICROBIOME

The leaf microbiome is considered an essential part of the plant that controls plant growth and development through transpiration and photosynthesis (Toju et al., 2019; Chouhan et al., 2021b). Studies conducted by different scientists have reported the abundance of specific phylum in the leaf microbiomes of Arabidopsis (Bai et al., 2015), tomato (Toju et al., 2019), rice (Roman-Reyna et al., 2020), tea (Cernava et al., 2019), etc. As a result, the leaf microbiome has the potential to modify the plant's functional qualities to protect it from changing climatic conditions. The study conducted by Bai et al. (2015) found that Gammaproteobacteria (*Pseudomonas*) were dominating leaves of Arabidopsis while Toju et al. (2019) in their study reported that members belonging to Alphaproteobacteria (like *Sphingomonas* and *Methylobacterium*) were found to be more dominating than other leaf-associated bacterial communities in the tomato plant. A similar study was reported by Wagner et al. (2020) that the mature leaf of maize was dominated by the species *Methylobacterium* and *Sphingomonas*. The phylum Proteobacteria has been found in the majority of the leaf (Wallace et al., 2018; Roman-Reyna et al., 2020). The study conducted by Roman-Reyna et al. (2020) used whole genome sequencing for the characterization of the leaf microbiome of 3,000 Rice Genomes Project. Bacterial genera like *Pseudomonas, Xanthomonas Burkholderia,* and *Mycoplasma* have been reported as the most dominating genera in the rice leaf microbiome. Among fungi basidiomycete (order Ustiaginales and

Tremellales) and ascomycete (order Plesporales and Capnodiales) were found to colonize tomato leaves. The study conducted by Toju et al. (2019) reported *Cladosporium, Dioszegia, Moesziomyces,* and *Hannaella* were colonizing tomato leaf microbiome. Another study reported *Dioszegia* (Basidiomycetes), belonging to Tremellales can play a role in microbe–microbe interaction webs of leaf-associated microbiome of cereals and *Arabidopsis* plants (Agler et al., 2016) (Figure 3.1).

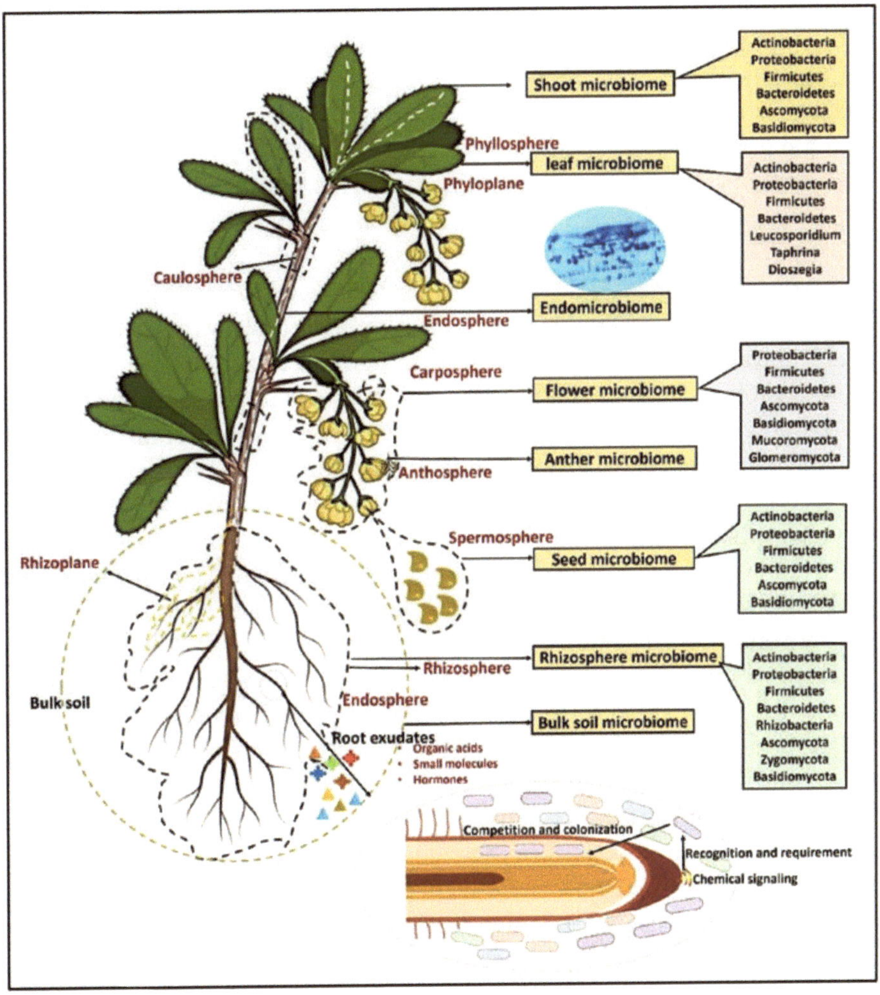

FIGURE 3.1 Diagrammatic depiction of the phytomicrobiome of different plant parts colonized by bacterial and fungal species.

3.4.2 SEED MICROBIOME

Seeds are the principal product of many crop plants, which are used for both human consumption and the start of the following crop cycle. As a result, for agricultural output, a healthy and high-quality seed is critical. Seeds, like other sections of a plant, are colonized by a diverse set of microbes that may have a variety of relationships with plants, including commensalism, interspecific cooperation, mutualism, and pathogenic associations that can distress the plant's health (Nicolaisen et al., 2014; Links et al., 2014). Different varieties of a crop could be colonized by the distinct microbiome on the seed surface or within the seed that may act as a reservoir for the next generation of the endophytic microbiome that assembles and shape the phytomicrobiome (Nelson, 2018). A wide range of microbial taxa can colonize in different seeds including phyla *Actinobacteria, Proteobacteria, Bacteroidetes,* and *Firmicutes* (Barret et al., 2015; Johnston-Monje et al., 2016). The study conducted by López-Ráez et al. (2017) reported that bacteria belonging to *Actinobacteria, Firmicutes*, and *Proteobacteria,* dominate in the seeds of alfalfa and *Cucurbita pepo* (Adam et al., 2018). The seed microbiome varies with different environmental conditions, types of soil, and plant species (Links et al., 2014; Barret et al., 2015; Nelson, 2018). According to Geisen et al. (2017), the Ascomycota group dominated in seed fungal endophytes, which are thought to operate as a source for germinating seeds. Although it has been suggested that components of seed microbiomes can be vertically transmitted from one generation to another (Vujanovic et al., 2019), the assembly process of seed microbiome remains poorly understood (Nelson, 2018). How seed microbiome gather and what ecological processes govern its association remains largely unknown. Functionally, seed-borne endophytes release a range of important substances such as enzymes phytohormones, and biocontrol agents, and other metabolic products that enhance plant performance and increase plant productivity under stress conditions (Shahzad et al., 2017; Camargo et al., 2019; Mukherjee et al., 2020).

The study conducted by Mukherjee et al. (2020) provided evidence that seed microbiome inoculation into the plants, can enhance the plant growth of chickpeas by introducing various plant growth-promoting (PGP) traits. This study suggests that seed microbiome have significant potential to mitigate abiotic and biotic stresses and thereby enhance crop productivity. A discovery indicates that flower inoculation may be used to change the composition of the seed microbiome, giving a possible method for testing scientific questions and optimizing this in the future to harness seed microbiomes for sustainable agriculture (Mitter et al., 2017). However, the seed microbiome, its metabolic capabilities, evolution, generational transmission and role in shaping phytomicrobiome is probably the least studied and should be the focus of future research to better understand their role in the establishment of seedlings and association of phytomicrobiome.

3.4.3 MICROBIOME ASSOCIATED WITH FLOWER

Plant microbiomes play critical roles in plant health and performance. Even though flowers are directly connected to reproductive success, little is known about their microbiomes and their possible relationship with pathogen attacks (Cui et al., 2021). Flowers are crucial for most plants' biological success, yet they are also the plant components most sensitive to biotic and abiotic stressors. Flower (anthosphere) microbiomes play an important role in the maintenance of crucial ecosystem services such as plant reproductive success and interactions with helpful insects (e.g., pollinators). However, we know significantly less about the microbial populations associated with flowers than we do about the phyllosphere microbiome. This is due in part to the nutrient-rich and high osmotic conditions in nectar, as well as the fact that the flower is an open habitat visited by a varied range of insects and other animals that transfer both diseases and plant probiotic components (Kim et al., 2019). The study conducted by Cui et al. (2021) studied apple flower microbiome under phytopathogen *Erwinia amylovora* infection. A previous study conducted by Steven et al. (2018) has shown that the stigma microbiome has limited diversity, even though specific lineages, mostly from the families *Pseudomonadaceae* and *Enterobacteriaceae*, are prominent (Table 3.1).

3.5 PHYTOMICROBIOME FOR IMPROVING AGRICULTURAL SUSTAINABILITY

For several decades, traditional farming methods have been primarily reliant on the use of artificial agrochemicals and enhanced plant cultivars, which have substantially increased crop production and food supply (Singh et al., 2018). Agriculture productivity, on the other hand, is declining structurally, and pesticide and plant breeding-induced productivity increases appear to have peaked (Trivedi et al., 2020). Furthermore, indiscriminate and disproportionate use of agrochemicals results in contaminated food and energy losses, as well as negative impact on the environment and human health

TABLE 3.1 Techniques Involved and the Communities Identified for Revealing the Potentials of Phytomicrobiome

Phytomicrobiome	Plant System	Location	Community Identified	Method Used	References
Rhizosphere microbiome	Soybean (*Glycine max*)	Brazil	Actinobacteria, Aquificae Cyanobacteria, Chloroflexi, Acidobacteria, Tenericutes, Chlamydiae, Chlorobi, Verrucomicrobia, and Deferribacteres	Roche 454 automated sequencer	Mendes et al. (2014)
	Soybean (*Glycine max*)	Japan	Proteobacteria, Actinobacteria, and Chloroflexi	454 pyrosequencing	Sugiyama et al. (2014)
	Sorghum (*Sorghum bicolor*)	South Africa	Cyanobacteria, Bacteroidetes, and Proteobacteria	Illumina sequencing	Kumar et al. (2021)
	Palash (*Butea monosperma*)	India	Actinobacteria, Proteobacteria, Firmicutes, and Bacteroidetes	Illumina sequencing	Vyas, Preeti, Kumar et al. (2019)
	Tomato (*Lycopersicon esculentum*)	Korea	Firmicutes and Actinobacteria	Illumina sequencing	Kwak et al. (2018); Lee et al. (2021)
	Banana (*Musa paradisiaca*)	China	Proteobacteria, Chloroflexi, Acidobacteria, Firmicutes, Gemmatimonadetes, Bacteroidetes, Actinobacteria, Verrucomicrobia Nitrospirae, and Planctomycetes	454-pyrose-quencing	Zhou et al. (2019)
	Banana (*Musa paradisiaca*)	Tanzania	Actinobacteria, Proteobacteria, and Bacteriodetes	Illumina sequencing	Kaushal et al. (2020)
	Spear grass	Africa	Bacterial: Actinobacteria and Alphaproteobacteria; and Fungal: *Aspergillus*, *Thielavia*, and *Curvularia*	Illumina sequencing	Marasco et al. (2018)
	Maize (*Zea mays*)	Raleigh	Acidobacteria, Chloroflexi, Verrucomicrobia, and Planctomycetes	Illumina sequencing	Peiffer et al. (2013)

TABLE 3.1 *(Continued)*

Phytomicrobiome	Plant System	Location	Community Identified	Method Used	References
Leaf microbiome	Rice	Philippines and China	*Pseudomonas*, *Xanthomonas Burkholderia*, and *Mycoplasma*	Whole genome sequencing	Roman-Reyna et al. (2020)
	Thale grass *Arabidopsis thaliana*	Germany	Gammaproteobacteria (*Pseudomonas*)	Whole genome sequencing	Bai et al. (2015)
	Tea (*Camellia sinensis*)	China	Actionobacteria, Proteobacteria, Bacteroidetes, Planctomycetes, and Verrucomicrobia	Illumina sequencing	Cernava et al. (2019)
	Tomato (*Lycopersicon esculentum*)	Japan	Alphaproteobacteria (like *Sphingomonas* and *Methylobacterium*)	Illumina sequencing	Toju et al. (2019)
	Maize (*Zea mays*)	USA	*Methylobacterium* and *Sphingomonas*	Illumina sequencing	Wagner et al. (2020)
Phyllosphere	Bean, soybean, and canola	USA	Firmicutes, Bacteroidetes, Thermi, and Chloroflexi	Illumina sequencing	Nuccio et al. (2020)
	Spinach	USA	Proteobacteria and Firmicutes	Pyrosequencing	Lopez-Velasco et al. (2011)
	Arabidopsis thaliana	Germany	Fungi: *Albugo* sp., *Dioszegia* sp., *Udeniomyces* sp. Bacteria: *Caulobacter* sp., *Flavobacterium*, and oomycetes symbionts	Illumina sequencing	Sánchez-Vallet et al. (2015)
Flower microbiome	Apple (*Malus* sp.)	USA	Enterobacteriaceae and Pseudomonadaceae	Illumina sequencing	Cui et al. (2021)
	Apple (*Malus* sp.)	Hamden	Pseudomonadaceae and Enterobacteriaceae	Illumina sequencing	Steven et al. (2018)

(e.g., heart-related problems, neurological disorders, endocrine disorders, chronic kidney diseases, cancer, female and male infertility) and food safety (Raj et al., 2021; Malla et al., 2022). Harnessing the phytomicrobiome is regarded as a promising developing strategy for avoiding these challenges and contributing to the objective of long-term gains in agricultural output (Singh et al., 2020). Agricultural production is essentially determined by two key factors: nutrient uptake and stress resistance. Changing the phytomicrobiome or its activity can improve both of these parameters. Agriculture microbial tools are already expanding quickly. According to current market research, the microbial amendment is one of the fastest-growing businesses, rising at a pace of 17% per year, which is 3.5 times faster than agrochemical markets and is expected to reach $12 billion by 2026.

Plant development and soil health are both aided by phytomicrobiomes (Smith et al., 2017; Dubey et al., 2019). A variety of microorganisms are known to influence plant resource allocation between root, shoot, flower, and seed, as well as to promote above and below-ground interactions with plant pests and other natural challenges (Schulz-Bohm et al., 2018). Furthermore, interactions between plants and microorganisms can alter the physicochemical and biological features of soil through co-production and organic compound alteration (Canarini et al., 2019). Some plants emit phytosiderophores, while bacteria secrete siderophores, which aid in iron sequestration from the soil (Ahmed & Holmström, 2014). Furthermore, these linkages are complex, and our present understanding of the physiological, molecular, abiotic and biotic components that control these interactions is limited. To harness them for agricultural production, a thorough knowledge of the relationships and implications of essential processes is required (Saleem & Moe, 2014). Many study researchers have described that PGPR, biostimulants, bio-inoculants, biofertilizers, and bio-control agents are potential agrochemical alternatives/supplements for increasing sustainable agricultural productivity (Ruzzi et al., 2016; Etminani & Harighi, 2018; Dubey et al., 2021, 2022). Several solutions are now on the market to boost N and P availability for crop uptakes (Sharma et al., 2013).

Culturable PGPR and endophytes have enormous potential to boost agricultural productivity and will continue to be the primary instrument for reducing the usage of chemical fertilizers for the foreseeable future (Barnawal et al., 2017; Dubey et al., 2020a, 2022). However, one of the primary limits of PGPR is the inconsistency of product efficacy. In other circumstances, the failure of PGPR in field effectiveness was ascribed to a lack of competitive advantage over natural soil microflora, an inability to colonize plant roots, and a lack of desirable activities. The primary difficulty with the effective nature of PGPR is that farmers do not apply it correctly as per the prescribed protocol in the field, as well as other concerns with the bioformulation synthesis in a non-suitable binding agent and their survival in diverse soil types (Jorjani et al., 2011; Chakraborty et al., 2013; Arora & Mishra, 2016).

Rhizosphere engineering is seen as a mid-term aim for *in situ* microbiome modification. Plant breeding and genetic engineering can alter the rhizosphere microbiome composition or activities that increase productivity (Mendes et al., 2013, 2019). Harnessing interactions between pathogens, root exudates, and related virulence factors, for example, might provide a significant way to redirect the rhizosphere microbiome (Dubey et al., 2020b; Kumar & Dubey, 2020; Yadav et al., 2020). Phytomicrobiomes can be genetically engineered to acquire useful features from other environmental bacteria (Ke et al., 2021). Endophytic strains and PGP microorganisms can be genetically engineered to convey desired functional features. Genetic alteration can also be used to improve microbial colonization of the host and/or give particular phenotypes (Vorholt et al., 2017). However, in addition to technological problems (such as transmitting multi-gene features), there are substantial social issues that must be addressed.

3.6 CONCLUSION AND FUTURE PERSPECTIVE

In recent years, the use of different microbes-based products such as biostimulants, biofertilizers, and bio-control agents has increased dramatically, although the success of this method has been variable. For long-term success and for better efficacy, continuous improvements in bioinoculant tools are needed, such as the use of consortia, synthetic community (as opposed to single isolate), endophytic microbes, and better formulation technologies. However, we believe that the future of the microbial approach rests in our capacity to manipulate the indigenous phytomicrobiome *in situ* to encourage the activity of beneficial bacteria. Overall, some progress has been made in the phytomicrobiome field in recent years, but many knowledge gaps remain, including: (i) an understanding microbial colonization process; (ii) characterization of biochemical and molecular signals that apprentice beneficial microbes in the rhizosphere and endosphere; and (iii) identification of host-associated core and hub microbiota and how they are influenced by agricultural practices. If these shortcomings are addressed systematically, phytomicrobiome techniques have the potential to give future tools for a transformative boost in agricultural production and environmental sustainability. In the future, phytomicrobiomes may prove to be more effective instruments as microbial inoculants and bio-controlling agents for plant growth and development under various environmental stresses.

ACKNOWLEDGMENTS

The authors would like to acknowledge the Department of Biotechnology (DBT), Government of India for partial financial assistance in the form of the project "Mapping and quantitative assessment of plant resources and its distribution in Madhya Pradesh, Central India" (Ref. No. BT/PR12899/NDB/39/506/2015 dt. 20/06/2017). Ashwani Kumar gratefully acknowledges DST-SERB for financial support obtained through a project grant of (CRG/2021/003696), New Delhi Govt of India.

KEYWORDS

- **agricultural sustainability**
- **leaf microbiome**
- **microbiome**
- **phytomicrobiome**
- **rhizosphere microbiome**
- **seed microbiome**
- **sustainable agriculture**

REFERENCES

Adam, E., Bernhart, M., Müller, H., et al., (2018). The *Cucurbita pepo* seed microbiome: Genotype-specific composition and implications for breeding. *Plant Soil, 422*, 35–49. https://doi.org/10.1007/s11104-016-3113-9.

Agler, M. T., Ruhe, J., Kroll, S., et al., (2016). Microbial hub taxa link host and abiotic factors to plant microbiome variation. *PLoS Biol., 14*. https://doi.org/10.1371/journal.pbio.1002352.

Ahmed, E., & Holmström, S. J. M., (2014). Siderophores in environmental research: Roles and applications. *Microb. Biotechnol., 7*, 196–208. https://doi.org/https://doi.org/10.1111/1751-7915.12117.

Anwar, S., Ali, B., & Sajid, I., (2016). Screening of rhizospheric actinomycetes for various in-vitro and in-vivo plant growth promoting (PGP) traits and for agroactive compounds. *Front Microbiol., 7*. https://doi.org/10.3389/fmicb.2016.01334.

Arora, N. K., & Mishra, J., (2016). Prospecting the roles of metabolites and additives in future bioformulations for sustainable agriculture. *Appl. Soil Ecol.*

Bai, Y., Müller, D. B., Srinivas, G., et al., (2015). Functional overlap of the *Arabidopsis* leaf and root microbiota. *Nature, 528*, 364–369. https://doi.org/10.1038/nature16192.

Barnawal, D., Bharti, N., Pandey, S. S., et al., (2017). Plant growth-promoting rhizobacteria enhance wheat salt and drought stress tolerance by altering endogenous phytohormone levels and TaCTR1/TaDREB2 expression. *Physiol. Plant, 161*, 502–514. https://doi.org/10.1111/ppl.12614.

Barret, M., Briand, M., Bonneau, S., et al., (2015). Emergence shapes the structure of the seed microbiota. *Appl. Environ. Microbiol., 81*, 1257—1266. https://doi.org/10.1128/aem.03722-14.

Berendsen, R. L., Pieterse, C. M. J., & Bakker, P. A. H. M., (2012). The rhizosphere microbiome and plant health. *Trends Plant Sci., 17*, 478–486.

Bhattacharyya, D., Duta, S., Yu, S. M., et al., (2018). Taxonomic and functional changes of bacterial communities in the rhizosphere of Kimchi cabbage after seed bacterization with *Proteus vulgaris* JBLS202. *Plant Pathol. J., 34*, 286–296. https://doi.org/10.5423/PPJ.OA.03.2018.0047.

Bulgarelli, D., Garrido-Oter, R., Münch, P. C., et al., (2015). Structure and function of the bacterial root microbiota in wild and domesticated barley. *Cell Host Microbe., 17*, 392–403. https://doi.org/10.1016/j.chom.2015.01.011.

Busby, P. E., Peay, K. G., & Newcombe, G., (2016). Common foliar fungi of *Populus trichocarpa* modify *Melampsora* rust disease severity. *New Phytol., 209*, 1681–1692. https://doi.org/10.1111/nph.13742.

Camargo, A. P., De Souza, R. S. C., De Britto, C. P., et al., (2019). Microbiomes of *Velloziaceae* from phosphorus-impoverished soils of the Campos rupestres, a biodiversity hotspot. *Sci. Data, 6*, 140. https://doi.org/10.1038/s41597-019-0141-3.

Canarini, A., Kaiser, C., Merchant, A., et al., (2019). Root exudation of primary metabolites: Mechanisms and their roles in plant responses to environmental stimuli. *Front Plant Sci., 10*, 157. https://doi.org/10.3389/fpls.2019.00157.

Cernava, T., Chen, X., Krug, L., et al., (2019). The tea leaf microbiome shows specific responses to chemical pesticides and biocontrol applications. *Sci. Total Environ., 667*, 33–40. https://doi.org/10.1016/j.scitotenv.2019.02.319.

Chakraborty, U., Chakraborty, B. N., Chakraborty, A. P., et al., (2013). Plant growth promoting rhizobacteria mediated improvement of health status of tea plants. *Indian J. Biotechnol.*

Chen, S., Waghmode, T. R., Sun, R., et al., (2019). Root-associated microbiomes of wheat under the combined effect of plant development and nitrogen fertilization. *Microbiome*. https://doi.org/10.1186/s40168-019-0750-2.

Chouhan, G. K., Mukherjee, A., Gaurav, A. K., et al., (2021a). Chapter 8 – plant-specific microbiome for environmental stress management: Issues and challenges. In: Verma, J. P., Macdonald, C. A., Gupta, V. K., & Podile, A. R., (eds.), *New and Future Developments in Microbial Biotechnology and Bioengineering* (pp. 69–89). Elsevier.

Chouhan, G. K., Verma, J. P., Jaiswal, D. K., et al., (2021b). Phytomicrobiome for promoting sustainable agriculture and food security: Opportunities, challenges, and solutions. *Microbiol. Res., 248,* 126763. https://doi.org/10.1016/j.micres.2021.126763.

Compant, S., Samad, A., Faist, H., & Sessitsch, A., (2019). A review on the plant microbiome: Ecology, functions, and emerging trends in microbial application. *J. Adv. Res., 19*, 29–37. https://doi.org/10.1016/j.jare.2019.03.004.

Cordero, J., De Freitas, J. R., & Germida, J. J., (2020). Bacterial microbiome associated with the rhizosphere and root interior of crops in Saskatchewan, Canada. *Can. J. Microbiol., 66*, 71–85. https://doi.org/10.1139/cjm-2019-0330.

Cui, Z., Huntley, R. B., Zeng, Q., & Steven, B., (2021). Temporal and spatial dynamics in the apple flower microbiome in the presence of the phytopathogen *Erwinia amylovora*. *ISME J., 15*, 318–329. https://doi.org/10.1038/s41396-020-00784-y.

Dhankher, O. P., & Foyer, C. H., (2018). Climate resilient crops for improving global food security and safety. *Plant Cell Environ.*

Dubey, A., Kumar, A., & Khan, M. L., (2020a). Role of biostimulants for enhancing abiotic stress tolerance in Fabaceae plants. In: *The Plant Family Fabaceae* (pp. 223–236).

Dubey, A., Kumar, A., Khan, M. L., & Payasi, D. K., (2022). Plant growth-promoting and bio-control activity of micrococcus luteus strain AKAD 3-5 isolated from the soybean (*Glycine max* (L.) Merr.) rhizosphere. *Open Microbiol J., 15*, 188–197. https://doi.org/10.2174/1874285802115010188.

Dubey, A., Malla, M. A., Khan, F., et al., (2019). Soil microbiome: A key player for conservation of soil health under changing climate. *Biodivers Conserv., 28*, 2405–2429. https://doi.org/10.1007/s10531-019-01760-5.

Dubey, A., Malla, M. A., Kumar, A., et al., (2020b). Plants endophytes: Unveiling hidden agenda for bioprospecting toward sustainable agriculture. *Crit. Rev. Biotechnol., 40*, 1210–1231.

Dubey, A., Saiyam, D., Kumar, A., et al., (2021). Bacterial root endophytes: Characterization of their competence and plant growth promotion in soybean (*Glycine max* (L.) merr.) under drought stress. *Int. J. Environ. Res. Public Health, 18*, 1–20. https://doi.org/10.3390/ijerph18030931.

Engelbrektson, A., Kunin, V., Engelbrektson, A., et al., (2012). Defining the core *Arabidopsis thaliana* root microbiome. *Nature*.

Etminani, F., & Harighi, B., (2018). Isolation and identification of endophytic bacteria with plant growth promoting activity and biocontrol potential from wild pistachio trees. *Plant Pathol. J.* https://doi.org/10.5423/PPJ.OA.07.2017.0158.

FAO, (2009). *How to Feed the World in 2050*? Insights from an Expert Meet FAO.

FAO, ITPS, GSBI, & SCBD, (2020). *State of Knowledge of Soil Biodiversity – Status, Challenges and Potentialities.* FAO.

Fasusi, O. A., Cruz, C., & Babalola, O. O., (2021). Agricultural sustainability: Microbial biofertilizers in rhizosphere management. *Agric., 11*.

Fitzpatrick, C. R., Copeland, J., Wang, P. W., et al., (2018). Assembly and ecological function of the root microbiome across angiosperm plant species. *Proc. Natl. Acad. Sci., 115*, E1157–E1165. https://doi.org/10.1073/pnas.1717617115.

Geisen, S., Kostenko, O., Cnossen, M. C., et al., (2017). Seed and root endophytic fungi in a range expanding and a related plant species. *Front. Microbiol., 8*, 1645. https://doi.org/10.3389/fmicb.2017.01645.

Ghaffari, M. R., Mirzaei, M., Ghabooli, M., et al., (2019). Root endophytic fungus *Piriformospora indica* improves drought stress adaptation in barley by metabolic and proteomic reprogramming. *Environ. Exp. Bot.* https://doi.org/10.1016/j.envexpbot.2018.10.002.

Gontia-Mishra, I., Sapre, S., Sharma, A., & Tiwari, S., (2016). Amelioration of drought tolerance in wheat by the interaction of plant growth-promoting rhizobacteria. *Plant Biol., 18*, 992–1000. https://doi.org/10.1111/plb.12505.

Johnston-Monje, D., Lundberg, D. S., Lazarovits, G., et al., (2016). Bacterial populations in juvenile maize rhizospheres originate from both seed and soil. *Plant Soil, 405*, 337–355. https://doi.org/10.1007/s11104-016-2826-0.

Jorjani, M., Heydari, A., Zamanizadeh, H. R., et al., (2011). Development of *Pseudomonas fluorescens* and *Bacillus coagulans* based bioformulations using organic and inorganic carriers and evaluation of their influence on growth parameters of sugar beet. *J. Biopestic., 4*, 180–185.

Kaushal, M., Mahuku, G., & Swennen, R., (2020). Metagenomic insights of the root colonizing microbiome associated with symptomatic and non-symptomatic bananas in *Fusarium* wilt infected fields. *Plants, 9*. https://doi.org/10.3390/plants9020263.

Ke, J., Wang, B., & Yoshikuni, Y., (2021). Microbiome engineering: Synthetic biology of plant-associated microbiomes in sustainable agriculture. *Trends Biotechnol., 39*, 244–261. https://doi.org/https://doi.org/10.1016/j.tibtech.2020.07.008.

Kim, D. R., Cho, G., Jeon, C. W., et al., (2019). A mutualistic interaction between *Streptomyces* bacteria, strawberry plants, and pollinating bees. *Nat. Commun., 10*, 4802. https://doi.org/10.1038/s41467-019-12785-3.

Krishna, S. B. N., Dubey, A., Malla, M. A., et al., (2019). Integrating microbiome network: Establishing linkages between plants, microbes, and human health. *Open. Microbiol. J., 13*, 330–342. https://doi.org/10.2174/1874285801913020330.

Kudoyarova, G. R., Melentiev, A. I., Martynenko, E. V., et al., (2014). Cytokinin producing bacteria stimulate amino acid deposition by wheat roots. *Plant Physiol. Biochem., 83*, 285–291. https://doi.org/https://doi.org/10.1016/j.plaphy.2014.08.015.

Kumar, A., & Dubey, A., (2020). Rhizosphere microbiome: Engineering bacterial competitiveness for enhancing crop production. *J. Adv. Res., 24*, 337–352.

Kumar, A., Dubey, A., Malla, M. A., & Dames, J., (2021). Pyrosequencing and phenotypic microarray to decipher bacterial community variation in *Sorghum bicolor* (L.) Moench rhizosphere. *Curr. Res. Microb. Sci.* https://doi.org/10.1016/j.crmicr.2021.100025.

Kwak, M. J., Kong, H. G., Choi, K., et al., (2018). Rhizosphere microbiome structure alters to enable wilt resistance in tomato. *Nat. Biotechnol., 36*, 1100–1109. https://doi.org/10.1038/nbt.4232.

Lakshmanan, V., Selvaraj, G., & Bais, H. P., (2014). Functional soil microbiome: Belowground solutions to an aboveground problem. *Plant Physiol., 166*, 689–700. https://doi.org/10.1104/pp.114.245811.

Lee, S. M., Kong, H. G., Song, G. C., & Ryu, C. M., (2021). Disruption of firmicutes and actinobacteria abundance in tomato rhizosphere causes the incidence of bacterial wilt disease. *ISME J., 15*, 330–347. https://doi.org/10.1038/s41396-020-00785-x.

Links, M. G., Demeke, T., Gräfenhan, T., et al., (2014). Simultaneous profiling of seed-associated bacteria and fungi reveals antagonistic interactions between microorganisms within a shared epiphytic microbiome on *Triticum* and *Brassica* seeds. *New Phytol., 202*, 542–553. https://doi.org/10.1111/nph.12693.

Liu, H., Macdonald, C. A., Cook, J., et al., (2019). An ecological loop: Host microbiomes across multitrophic interactions. *Trends Ecol. Evol., 34,* 1118–1130. https://doi.org/https://doi.org/10.1016/j.tree.2019.07.011.

López-Ráez, J. A., Shirasu, K., & Foo, E., (2017). Strigolactones in plant interactions with beneficial and detrimental organisms: The yin and yang. *Trends Plant Sci., 22*, 527–537.

Lopez-Velasco, G., Welbaum, G. E., Boyer, R. R., et al., (2011). Changes in spinach phylloepiphytic bacteria communities following minimal processing and refrigerated storage described using pyrosequencing of 16S rRNA amplicons. *J. Appl. Microbiol., 110*, 1203–1214. https://doi.org/10.1111/j.1365-2672.2011.04969.x.

Mahapatra, S., Chandra, B., Viswavidyalaya, K., et al., (2020). *Phytobiomes: Current Insights and Future Vistas*. Springer Singapore, Singapore.

Malla, M. A., Dubey, A., Kumar, A., et al., (2019). *Exploring the Human Microbiome: The Potential Future Role of Next-Generation Sequencing in Disease Diagnosis and Treatment*, 9. https://doi.org/10.3389/fimmu.2018.02868.

Malla, M. A., Dubey, A., Raj, A., et al., (2022). Emerging frontiers in microbe-mediated pesticide remediation: Unveiling role of omics and in silico approaches in engineered environment. *Environ. Pollut., 299*, 118851. https://doi.org/10.1016/j.envpol.2022.118851.

Malla, M. A., Dubey, A., Yadav, S., et al., (2018). Understanding and designing the strategies for the microbe-mediated remediation of environmental contaminants using omics approaches. *Front. Microbiol., 9.* https://doi.org/10.3389/fmicb.2018.01132.

Marasco, R., Mosqueira, M. J., Fusi, M., et al., (2018). Rhizosheath microbial community assembly of sympatric desert speargrasses is independent of the plant host. *Microbiome, 6*, 215. https://doi.org/10.1186/s40168-018-0597-y.

Maxton, A., Singh, P., & Masih, S. A., (2018). ACC deaminase-producing bacteria mediated drought and salt tolerance in *Capsicum annuum*. *J. Plant Nutr., 41*, 574–583. https://doi.org/10.1080/01904167.2017.1392574.

Mendes, L. W., De Chaves, M. G., Fonseca, M. D. C., et al., (2019). Resistance breeding of common bean shapes the physiology of the rhizosphere microbiome. *Front Microbiol.* https://doi.org/10.3389/fmicb.2019.02252.

Mendes, L. W., Kuramae, E. E., Navarrete, A. A., et al., (2014). Taxonomical and functional microbial community selection in soybean rhizosphere. *ISME J., 8*, 1577–1587. https://doi.org/10.1038/ismej.2014.17.

Mendes, R., Garbeva, P., & Raaijmakers, J. M., (2013). The rhizosphere microbiome: Significance of plant beneficial, plant pathogenic, and human pathogenic microorganisms. *FEMS Microbiol. Rev., 37*, 634–663. https://doi.org/10.1111/1574-6976.12028.

Miransari, M. (Ed.). (2016). Abiotic and Biotic Stresses in Soybean Production: Soybean Production Volume 1 (Vol. 1). Academic press. https://doi.org/10.1016/C2014-0-00087-1.

Mitter, B., Pfaffenbichler, N., Flavell, R., et al., (2017). A new approach to modify plant microbiomes and traits by introducing beneficial bacteria at flowering into progeny seeds. *Front. Microbiol., 8*, 11. https://doi.org/10.3389/fmicb.2017.00011.

Mukherjee, A., Singh, B. K., & Verma, J. P., (2020). Harnessing chickpea (*Cicer arietinum* L.) seed endophytes for enhancing plant growth attributes and bio-controlling against *Fusarium* sp. *Microbiol. Res., 237*, 126469. https://doi.org/10.1016/j.micres.2020.126469.

Nelson, E. B., (2018). The seed microbiome: Origins, interactions, and impacts. *Plant Soil, 422*, 7–34. https://doi.org/10.1007/s11104-017-3289-7.

Nicolaisen, M., Justesen, A. F., Knorr, K., et al., (2014). Fungal communities in wheat grain show significant co-existence patterns among species. *Fungal Ecol., 11*, 145–153. https://doi.org/https://doi.org/10.1016/j.funeco.2014.06.002.

Nuccio, E. E., Starr, E., Karaoz, U., et al., (2020). Niche differentiation is spatially and temporally regulated in the rhizosphere. *ISME J., 14,* 999–1014. https://doi.org/10.1038/s41396-019-0582-x.

Oliveira, I. J., Fontes, J. R. A., Pereira, B. F. F., & Muniz, A. W., (2018). Inoculation with *Azospirillum brasiliense* increases maize yield. *Chem. Biol. Technol. Agric.* https://doi.org/10.1186/s40538-018-0118-z.

Peiffer, J. A., Spor, A., Koren, O., et al., (2013). Diversity and heritability of the maize rhizosphere microbiome under field conditions. *Proc. Natl. Acad. Sci. U S A., 110*, 6548–6553. https://doi.org/10.1073/pnas.1302837110.

Pétriacq, P., Williams, A., Cotton, A., et al., (2017). Metabolite profiling of non-sterile rhizosphere soil. *Plant J., 92*, 147–162. https://doi.org/10.1111/tpj.13639.

Prudent, M., Salon, C., Souleimanov, A., et al., (2015). Soybean is less impacted by water stress using *Bradyrhizobium japonicum* and thuricin-17 from *Bacillus thuringiensis*. *Agron. Sustain. Dev., 35*, 749–757. https://doi.org/10.1007/s13593-014-0256-z.

Quiza, L., St-Arnaud, M., & Yergeau, E., (2015). Harnessing phytomicrobiome signaling for rhizosphere microbiome engineering. *Front. Plant Sci., 6*, 1–11. https://doi.org/10.3389/fpls.2015.00507.

Raj, A., Kumar, A., & Dames, J. F., (2021). Tapping the role of microbial biosurfactants in pesticide remediation: An eco-friendly approach for environmental sustainability. *Front. Microbiol., 12.* https://doi.org/10.3389/fmicb.2021.791723.

Rascovan, N., Carbonetto, B., Perrig, D., et al., (2016a). Integrated analysis of root microbiomes of soybean and wheat from agricultural fields. *Sci. Rep.* https://doi.org/10.1038/srep28084.

Rascovan, N., Carbonetto, B., Perrig, D., et al., (2016b). Integrated analysis of root microbiomes of soybean and wheat from agricultural fields. *Nat. Publ. Gr.,* 1–12. https://doi.org/10.1038/srep28084.

Roman-Reyna, V., Pinili, D., Borja, F. N., et al., (2020). Characterization of the leaf microbiome from whole-genome sequencing data of the 3000 rice genomes project. *Rice (N Y), 13*, 72. https://doi.org/10.1186/s12284-020-00432-1.

Ruzzi, M., & Aroca, R. (2015). Plant growth-promoting rhizobacteria act as biostimulants in horticulture. *Scientia Horticulturae, 196*, 124–134. https://doi.org/10.1016/j.scienta.2015.08.042.

Saleem, M., & Moe, L. A., (2014). Multitrophic microbial interactions for eco- and agro-biotechnological processes: Theory and practice. *Trends Biotechnol., 32*, 529–537. https://doi.org/https://doi.org/10.1016/j.tibtech.2014.08.002.

Sánchez-Vallet, A., Mesters, J. R., & Thomma, B. P. H. J., (2015). The battle for chitin recognition in plant-microbe interactions. *FEMS Microbiol. Rev., 39*,171–183.

Sandhya, V., Shrivastava, M., Ali, S. Z., & Sai Shiva, K. P. V., (2017). Endophytes from maize with plant growth promotion and biocontrol activity under drought stress. *Russ. Agric. Sci.* https://doi.org/10.3103/s1068367417010165.

Schreiter, S., Ding, G. C., Heuer, H., et al., (2014). Effect of the soil type on the microbiome in the rhizosphere of field-grown lettuce. *Front. Microbiol., 5.* https://doi.org/10.3389/fmicb.2014.00144.

Schulz-Bohm, K., Gerards, S., Hundscheid, M., et al., (2018). Calling from distance: Attraction of soil bacteria by plant root volatiles. *ISME J., 12*, 1252–1262. https://doi.org/10.1038/s41396-017-0035-3.

Shahzad, R., Waqas, M., Khan, A. L., et al., (2017). Indoleacetic acid production and plant growth promoting potential of bacterial endophytes isolated from rice (*Oryza sativa* L.) seeds. *Acta Biol. Hung., 68*, 175–186. https://doi.org/10.1556/018.68.2017.2.5.

Sharma, S. B., Sayyed, R. Z., Trivedi, M. H., & Gobi, T. A., (2013). *Phosphate Solubilizing Microbes: Sustainable Approach for Managing Phosphorus Deficiency in Agricultural Soils*. Springerplus.

Singh, S., Singh, H., Rout, B., et al., (2020). The new science of metagenomics: Revealing the secrets of microbial physiology. In: *Metagenomics: Techniques, Applications, Challenges and Opportunities* (pp. 3–22). Springer Singapore.

Singh, V. K., Singh, A. K., Singh, P. P., & Kumar, A., (2018). Interaction of plant growth promoting bacteria with tomato under abiotic stress: A review. *Agric. Ecosyst. Environ.*

Smith, D. L., Gravel, V., & Yergeau, E., (2017). Editorial: Signaling in the phytomicrobiome. *Front. Plant Sci., 8*.

Steven, B., Huntley, R. B., & Zeng, Q., (2018). The influence of flower anatomy and apple cultivar on the apple flower phytobiome. *Phytobiomes J., 2*, 171–179. https://doi.org/10.1094/PBIOMES-03-18-0015-R.

Sugiyama, A., Ueda, Y., Zushi, T., et al., (2014). Changes in the bacterial community of soybean rhizospheres during growth in the field. *PLoS One, 9*, 1–9. https://doi.org/10.1371/journal.pone.0100709.

Timmusk, S., Behers, L., Biocenter, U., et al., (2012). Enhancing drought tolerance in crops. *Irrig. Drain Syst. Eng.*, 1–5.

Toju, H., Okayasu, K., & Notaguchi, M., (2019). Leaf-associated microbiomes of grafted tomato plants. *Sci. Rep., 9*, 1787. https://doi.org/10.1038/s41598-018-38344-2.

Trivedi, P., Leach, J. E., Tringe, S. G., et al., (2020). Plant-microbiome interactions: From community assembly to plant health. *Nat. Rev. Microbiol., 18*, 607–621. https://doi.org/10.1038/s41579-020-0412-1.

Vardharajula, S., Ali, S. Z., Grover, M., et al., (2011). Drought-tolerant plant growth promoting *Bacillus* spp.: Effect on growth, osmolytes, and antioxidant status of maize under drought stress. *J. Plant Interact., 6*, 1–14. https://doi.org/10.1080/17429145.2010.535178.

Vorholt, J. A., Vogel, C., Carlström, C. I., & Müller, D. B., (2017). Establishing causality: Opportunities of synthetic communities for plant microbiome research. *Cell Host Microbe. 22*, 142–155.

Vujanovic, V., Islam, M. N., & Daida, P., (2019). Transgenerational role of seed mycobiome – an endosymbiotic fungal composition as a prerequisite to stress resilience and adaptive phenotypes in *Triticum*. *Sci. Rep., 9*, 18483. https://doi.org/10.1038/s41598-019-54328-2.

Vyas, P., Kumar, A., Malla, M. A., & Dubey, A., (2019). Taxonomic and functional annotation of termite degraded *Butea monosperma* (Lam.) Kuntze (flame of the forest). *Open Microbiol. J. 13*, 154–163. https://doi.org/10.2174/1874285801913010154.

Wagner, M. R., Busby, P. E., & Balint-Kurti, P., (2020). Analysis of leaf microbiome composition of near-isogenic maize lines differing in broad-spectrum disease resistance. *New Phytol., 225*, 2152–2165. https://doi.org/10.1111/nph.16284.

Wallace, J. G., Kremling, K. A., Kovar, L. L., & Buckler, E. S., (2018). Quantitative genetics of the maize leaf microbiome. *Phytobiomes J., 2*, 208–224. https://doi.org/10.1094/PBIOMES-02-18-0008-R.

Wolinska, A., Kuzniar, A., & Galazka, A., (2020). Biodiversity in the rhizosphere of selected winter wheat (*Triticum aestivum* L.) cultivars-genetic and catabolic fingerprinting. *Agronomy*. https://doi.org/10.3390/agronomy10070953.

Yadav, V. K., Raghav, M., Sharma, S. K., & Bhagat, N., (2020). Rhizobacteriome: Promising candidate for conferring drought tolerance in crops. *J. Pure Appl. Microbiol., 14*, 73–92.

Yeoh, Y. K., Paungfoo-Lonhienne, C., Dennis, P. G., et al., (2016). The core root microbiome of sugarcanes cultivated under varying nitrogen fertilizer application. *Environ. Microbiol., 18*, 1338–1351. https://doi.org/https://doi.org/10.1111/1462-2920.12925.

Yergeau, E., Bell, T. H., Champagne, J., et al., (2015). Transplanting soil microbiomes leads to lasting effects on willow growth, but not on the rhizosphere microbiome. *Front. Microbiol., 6,* 1436.

Zarraonaindia, I., Owens, S. M., Weisenhorn, P., et al., (2015). The soil microbiome influences grapevine-associated microbiota. *MBio*. https://doi.org/10.1128/mBio.02527-14.

Zhou, D., Jing, T., Chen, Y., et al., (2019). Deciphering microbial diversity associated with *Fusarium* wilt-diseased and disease-free banana rhizosphere soil. *BMC Microbiol., 19*. https://doi.org/10.1186/s12866-019-1531-6.

CHAPTER 4

Role of Metabolomics in Bioremediation for Sustainable Crop Production

AMAN RAJ,[1] ANAMIKA DUBEY,[1] MUNEER AHMAD MALLA,[1] and ASHWANI KUMAR[1]

[1]Metagenomics and Secretomics Research Laboratory, Department of Botany, Dr. Harisingh Gour Vishwavidyalaya (A Central University), Sagar, Madhya Pradesh, India

[2]Metagenomics and Secretomics Research Laboratory, Department of Botany, University of Allahabad (A Central University), Prayagraj, Uttar Pradesh, India

ABSTRACT

The world's hunger index is alarming, with more than 50% of the world population being undernourished, and the changing world climate attracts a diversity of pests, thus affecting crop productivity. Various chemical fertilizers and pesticides are being used to deal with these pests and pathogens and increase productivity on the limited land areas to feed the burgeoning human population. These agrochemicals solve the concern of dealing with pests, but most of these compounds are carcinogenic and have a long half-life, thus contaminating the diverse environmental bodies. Apart from these fertilizers and pesticides, various other anthropogenic activities lead to the discharge of diverse types of xenobiotics in the environment, such as heavy metals, pharmaceuticals, daily care products, oil spills, etc. Microbes residing in soil bodies play an important role in dealing with these agrochemicals and other xenobiotics as they utilize these chemicals as their sole source of carbon, thus making the soil contaminant free. The emergence of Omics technologies plays a vital role in the sustainable remediation of these pollutants. Among

Advances in Plant Microbiome Research for Climate Resilient Agriculture: Toward Sustainable Farming. Ashwani Kumar (Ed.)

all the omics technologies, metabolomics is the key tool in identifying the toxic compound, its degradation product, the biological molecule involved in degradation, and the enzymes and chemicals secreted by the microbes to degrade the agrochemical pollutants. This chapter discusses the role of various metabolomics tools such as GC-MS, LC-MS, and NMR for the sustainable remediation of toxic pollutants from the environment. Moreover, this chapter discusses different web sources and software for metabolomics data interpretation.

4.1 INTRODUCTION

Soil contamination and ecological degradation are worldwide issues caused by human and natural factors (Muneer et al., 2018; Raj et al., 2021). The world's population is growing at an alarming rate, which has accelerated industrialization and contemporary agricultural methods such as the widespread use of chemical fertilizers to fulfill the worldwide hunger need (Rohr et al., 2016). Anthropogenic contamination is mostly caused by industrialization and urbanization, while the use of agrochemicals for greater agricultural output has resulted in the dispersion and build-up of environmental pollutants (Prihandiani et al., 2021). Soil-bound pesticide residues accumulate 1/3 of total agrochemicals, or around 1 kg/hectare each year in the agricultural field (Boesten, 2016). It is vital to identify sustainable solutions to degrade environmental toxins to alleviate the problem of agrochemical build-up and environmental deterioration. To protect both the environment and human life, it's critical to understand and recognize the factors that influence the harmful agrochemicals' environmental fate, notably their degradation, which is the only long-term and cost-effective approach to eliminate them from the environment (Dubey et al., 2019; Kumar et al., 2021; Raj et al., 2021).

Bioremediation is a method of removing hazardous chemicals such as pesticides, heavy metals, oil spills, pharmaceutical drugs, and other types of xenobiotics that degrades the natural quality of soil and water bodies by combining microbial metabolism with optimal ecological parameters (Raj et al., 2021; Malla et al., 2021). Metabolites regulate microbial physiology under constant changes in the environment. Because they use xenobiotic pollutants to meet their nutritional needs, microorganisms in the soil operate as a nature reserve with strong xenobiotic degradation potential (Kumar et al., 2021). Metabolomics is a postgenomic approach that has been utilized to comprehend better diverse bioactive investigations that have been the focus of the study.

The metabolome is defined as the collection of low molecular weight molecules generated by biological entities. It investigates metabolites at a large scale in cells, tissues, organs, and organisms. Instrumental evaluation is one of the most crucial phases in metabolite detection and consequent pathway analysis. HPLC (high-performance liquid chromatography), NMR (nuclear magnetic resonance), and MS (mass spectrometry) are some of the advanced tools that can identify a huge variety of metabolites. NMR is a quasi-technique for obtaining detailed information on metabolite structures in suspension. Because no segregation or functionalization of metabolites is required, it has a significant advantage in the study of functional groups, volatile chemicals, sugar, and amino acids. In a quantitative ^{1}H NMR study, the accretion time of a high-field NMR spectrometer is significantly faster than that of GC or LC-MS, and the accuracy and consistency have been proved to be about 1% (Burton et al., 2005; Zhang et al., 2021). NMR is a heavy-duty, dynamic and extremely powerful tool for toxicokinetic evaluations of environmental species from polluted fields because of these features (Melvin et al., 2018). However, there runs a trade-off concept in every aspect, and no machine has 100% efficiency, and the same is the case with NMR. Low-specificity, non-selectivity, and redundant resonances are some of the downsides of NMR, which makes it quite not suitable for target-based metabolomics (Emwas, 2015). Two-dimensional sampling techniques can help to overcome the problems of redundant resonances.

However, NMR combined with MS is the best way to evaluate the metabolites from the environmental samples (Ludwig et al., 2010; Zhang et al., 2021). MS-based analyzers, which are constantly evolving, offer an advantage over NMR in that they can qualitatively or quantitatively profile metabolites from a complex mixture with excellent specificity and very high sensitivity (Villas-Bôas et al., 2005; Dettmer et al., 2007; Emwas, 2015). Among MS techniques, orbitrap, and TOF (time of flight) techniques are extensively used for xenobiotic metabolomics (Kim & Heyman, 2018). The MS technique can be used by direct infusion method or can be integrated with other chromatographic techniques such as LC-MS (liquid chromatography-mass spectrometry) or GC-MS (gas chromatography-mass spectrometry) to obtain the best results (Pöhö et al., 2019). Ambient organisms that cope with environmental stresses have been intensively researched using MS-based metabolite detection and combined chromatographic separation of metabolites (Zhang et al., 2021). After functionalization, GC-MS is used to analyze volatile and thermally stable chemicals with detection limits as low as nmol/pmol and great analytical repeatability. It works well with tiny molecules, including chemical compounds, amines, sugars, proteins, and fatty acid metabolism. It also outperforms NMR and LC-MS in cheap instrumentation costs (Dunn & Ellis, 2005; Aliferis & Chrysayi-Tokousbalides, 2011; Kostopoulou et al., 2020). Due to the tiny porosity in the column, which greatly reduces the time for analysis while retaining excellent separation efficiency, ultrahigh-pressure liquid chromatography-mass spectrometry (UPLC-MS) is frequently used in high-throughput metabolomics investigations.

LC-MS, on the other hand, has attracted attention in recent metabolomics investigations since it does not need sample functionalization and can analyze polar, quasi-polar, and non-polar metabolites (Núñez et al., 2012). The purpose of this chapter is to discuss the major environmental toxins, as well as the uses and significance of metabolomics in microbial bioremediation of environmental contaminants employing high-throughput instruments such as GC-MS, LC-MS, and NMR (Figure 4.1).

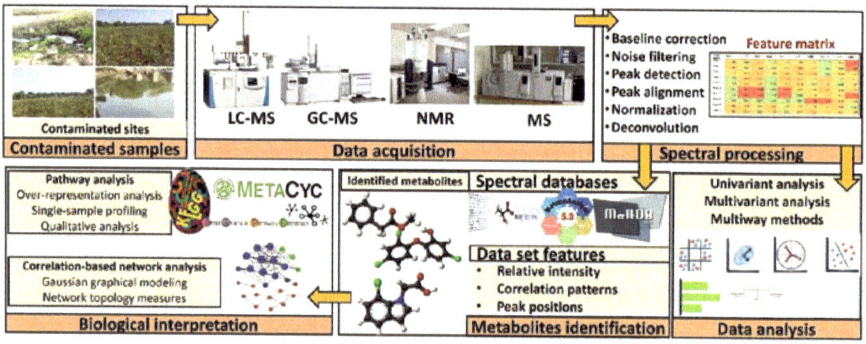

FIGURE 4.1 Workflow for untargeted metabolomic investigations.

4.2 METABOLOMICS

Metabolomics is a powerful method for understanding systems biology, including not just physicochemical analysis and data mining but also experimentation design and implementation, characterization of samples, pre-processing data analyzes, and biological interpretations (Aliferis & Chrysayi-Tokousbalides, 2011). In other words, the study of the metabolome in biological organisms is known as metabolomics. It concentrates on low molecular weight compounds, such as primary and secondary metabolites (SMs), which constitute the metabolome. Low molecular weight substances are critical for obtaining a physiological viewpoint and behavioral identity of organisms in their surroundings. The goal of metabolomics is to determine how different metabolites react to different stresses and how that relates to the individual's phenotype (Matich et al., 2019). Robinson and Pauling authored the first article on metabolomics in 1971 (Pauling et al., 1971), and since then, metabolomics approaches have been used in many fields such as bioremediation research, biomarker discovery, medication development, and environmental investigations. Nicholson and his colleagues examined clinical samples' first profiling in the 1980s, using high-throughput screening methods such as GC-MS and NMR for metabolic profiling. However, the term *metabolomics* came into existence quite later, and Oliver and his associates coined the word "metabolomics" in 1998 while exploring yeast physiology (Jeevanandam & Osborne, 2021). The decision of the appropriate facility that will be utilized to analyze samples is a crucial stage in metabolomics. The decision is based on various considerations, including the physicochemical features and chemical properties of the materials, the study's goal, and the technology available. Many analytical types of equipment for metabolomics have been invented, despite that, the most often used metabolomics analyzers are NMR and MS based analyzers (Misra et al., 2019; Matich et al., 2019; Misra, 2019).

4.2.1 NUCLEAR MAGNETIC RESONANCE (NMR)-BASED METABOLOMICS

In 1974, NMR was employed for the first time to study the metabolome (Hoult et al., 1974), and it has been about 50 years for NMR spectroscopy to enhance the knowledge and understanding of different physiological processes and their metabolism (Wishart, 2019). It provides quantifying ability and sample handling convenience, allowing metabolite extraction stages to be skipped. It's also a very accurate and non-destructive method. However, this method necessitates large concentrations, often in the millimolar range, and presents difficulties in automation and high throughput studies (Larive et al., 2015; Markley et al., 2017). NMR investigations can vary, including the nuclei used (e.g., ^{15}N, ^{1}H, ^{31}P, and ^{13}C) and the degree of synchronization provided. There exist different categories of NMR for different purposes and is divided into 1D-NMR spectroscopy and 2D-NMR spectroscopy. Since 1D-NMR (^{1}H NMR) makes component identification particularly difficult, especially when evaluating crude samples without preliminary separation, 2-D NMR spectroscopy such as total correlation spectroscopy (TOCSY), and correlation spectroscopy (COSY) is used for chemical analysis of the desired samples (Aliferis & Chrysayi-Tokousbalides, 2011; Emwas et al., 2019). NMR is the only metabolomic approach that is not limited to tissue extract or biofluid assessment since it is the best instrument for studying live tissues, organs, and other semi-solid/solid materials using the potential of ssNMR (solid-state NMR), and MAS-NMR (magic-angle spinning-NMR).

Furthermore, NMR enables users to investigate chemistry more than any other technology. It enables users to examine complete compounds at the atomic level, revealing not just ^{1}H atoms but also many other types of atoms (^{15}N, ^{13}C) or physiologically reactive groups, such as phosphate atoms. Here we have briefly discussed the role of each type of NMR in metabolomics studies.

4.2.2 ^{1}H NMR SPECTROSCOPY FOR METABOLOMICS RESEARCH

^{1}H NMR spectroscopy is the most routinely used method for metabolomics research, as ^{1}H atoms are found in nearly every organic component, including every known metabolite. Furthermore, 1D-^{1}H NMR spectroscopy is a highly automated, incredibly reliable, and very quick approach for metabolomics research since the spectrum acquisition time is as low as a few minutes, and many samples may be loaded and withdrawn continuously for days or even weeks utilizing sophisticated robotic instruments (Emwas et al., 2019). Aside from that, the approach can detect and analyze well over 50–100 metabolites through any biofluid or tissue sample in a single run (Hao et al., 2012). Furthermore, some public databases facilitate metabolite spectrum analysis by maintaining a massive quantity of data pertaining to spectral data for diverse metabolites. Many programs have recently been developed that can analyze, interpret, and quantify ^{1}H NMR spectra automatically in minutes or seconds. Because of these benefits in terms of repeatability and speed, ^{1}H NMR is the method of choice for large-scale metabolomics research involving thousands of samples (Tardivel et al., 2017; Cañueto et al., 2018). Despite its numerous advantages, ^{1}H NMR spectroscopy has certain drawbacks, such as a short chemical shift window that causes clashing of peaks, leading to higher ambiguity in metabolite detection and quantification. However, in 1D ^{1}H-NMR investigations, spectral resolution may be enhanced by employing bigger magnets or by utilizing 2D excitation techniques like TOCSY NMR experiments, however, collecting 2D spectra takes a long time (Cloarec et al., 2005; Rahman & Choudhary, 2015).

4.2.3 ^{13}C NMR SPECTROSCOPY FOR METABOLOMICS RESEARCH

When compared to 1H NMR spectroscopy, ^{13}C NMR spectroscopy has narrower line widths and a wider chemical shift dispersion. ^{13}C NMR Spectroscopy is widely utilized in isotope tracking and tracing experiments, which aid in the direct identification of the carbon source in various biochemical pathways and the elucidation of the pathway's whole chemistry (Schätzlein et al., 2018; Jang et al., 2018). This has led to its widespread use in NMR and MS-based metabolomics (Fluxomics) research (Heux et al., 2017). Fluxomics is a sub-discipline of metabolomics that studies internal metabolic outflows in live cells (Muneer et al., 2018). The procedure entails inserting isotopically amplified ^{13}C molecules into live cells and then measuring bioactivity by monitoring and analyzing ^{13}C isotope propagation along with the metabolic network under investigation (Zhang et al., 2014). In such cell-based NMR investigations, 1D-^{13}C direct detection or indirect ^{13}C detection employing 2D-NMR methods like ^{1}H-^{13}C HSQC (heteronuclear single quantum correlation spectroscopy) are frequently used (Wishart, 2019). The ^{13}C NMR has several limitations, including a low natural abundance of ^{13}C signals and a limited sensitivity to the ^{13}C nucleus. However, utilizing the DEPT (distortionless enhancement by polarization transfer) experiment, this restriction may be circumvented by increasing the ^{13}C signal strength by a factor of four (Markley et al., 2017). ^{31}P NMR spectroscopy and ^{15}N NMR spectroscopy are not commonly used in metabolomic studies due to the low abundance of nitrogen and phosphorus atoms (Aliferis & Chrysayi-Tokousbalides, 2011; Emwas et al., 2019).

4.2.4 COSY (CORRELATION SPECTROSCOPY) AND TOCSY (TOTAL CORRELATION SPECTROSCOPY) FOR METABOLOMICS

The simplest of all 2D-NMR experiments is the single, relayed COSY experiment. It is commonly used for chemical characterization and structural elucidation as it offers information on homonuclear correlations between linked nuclei (^{1}H-^{1}H) (Macura et al., 1983; Kono, 2013). The COSY technique is extremely advantageous for metabolomics studies since it is very simple, rapid, uncomplicated to execute, and easy to understand. Like other 2D NMR experiments, COSY studies allow for the identification of both known and unidentified metabolites, whereas 1D NMR experiments are mostly used to identify known metabolites (Blasco et al., 2010; Sekiyama et al., 2011).

TOCSY is a variation of the COSY technique in which the chemical shift of one nucleus is connected with the chemical displacement of other nuclei in a compound's total spin system. The cross-peaks are seen in the TOCSY spectrum not just for short-range coupled protons, but also for protons linked by a network of scalar couplings (Cloarec et al., 2005). While the 2D TOCSY experiment takes a long time to complete, the 1D TOCSY is much faster and generates a more unambiguous 1D NMR spectrum that is easier to interpret (Crook & Powers, 2020). When there is a lot of overlap in the NMR spectrum or you want to quantify highly overlapped metabolite species, the 1D TOCSY experiment comes in handy. TOCSY studies with selective excitation are another type of TOCSY technique that may be utilized to overcome the spectrum overlapping problem and help in metabolite detection (Sandusky et al., 2011; Lucio-Gutiérrez et al., 2019).

4.2.5 MASS SPECTROMETRY (MS)-BASED METABOLOMICS

Mass spectroscopy is the most often used analytical technology for metabolomics due to its excellent specificity, low detection limits, versatility for high-throughput analysis, and wide application possibilities (Aretz & Meierhofer, 2016). Different researchers have explored the feasibility of different MS-based analytical systems for metabolomics investigations from time to time (Emwas, 2015; Baidoo & Teixeira Benites, 2019; Misra, 2019; Al-Zubaidi et al., 2021). MS-based methods can provide precise quantitation using standards or other target molecules such as intermediaries and relative metabolite levels between samples (Hoffmann Edmond de, 2007). One of the most notable benefits of MS for metabolomics is the integration of MS detectors with chromatographic separation (such as LC-MS, GC-MS, and CE-MS) in conjunction with existing mass spectra libraries (Aliferis & Chrysayi-Tokousbalides, 2011; Matich et al., 2019). Liquid chromatography (LC) is used to determine variations in the zeta potential of non-volatile chemicals, whereas gas chromatography (GC) is used to determine the boiling temperatures of volatile compounds, and capillary electrophoresis (CE) is used to determine the ionic mobility of charged ions.

Mass spectrometry-based methods combined with LC (LC-MS) and GC (GC-MS) are by far the most widely utilized because they can cover a large component of the metabolome (Gross, 2011a). Due to the excellent chromatographic resolution, reliable retention periods, and convenience of use, GC-MS is preferable to LC-MS. It also has a higher throughput than other approaches as it demands less care (Kuehnbaum & Britz-Mckibbin,

2013; Matich et al., 2019). Moreover, GC is frequently used in conjunction with hard ionization sources like electron impact ionization (EI), allowing for in-source disintegration and determination of the molecular ion using the vast GC-MS databases. On the other hand, the utilization of GC-MS is restricted to volatile analytes. However, non-volatile substances like amino acid residues, polysaccharides, and organic acids can be investigated by GC-MS, but they must be derivatized first (Baidoo & Teixeira Benites, 2019). Capillary GC columns are used in most metabolomics research since they give enhanced specificity and excellent separation efficiency. The stationary phase (internal coating) is covalently attached to the inside surface of the GC column and has a diameter of 0.1–5 μm in general. The most widely utilized stationary phases for GC are methyl-silicon and polyethene glycol. Helium gas is used to transport analytes through the GC column since it is inert and compatible with most sensors (Huebschmann, 2015).

The introduction and use of time-of-flight (TOF) analysis tools in conjunction with GC have considerably decreased the time needed for complicated mixture investigations while improving mass accuracy (Aliferis & Chrysayi-Tokousbalides, 2011). Two-dimensional gas chromatographic separation can boost GC efficiency further (Kuehnbaum & Britz-Mckibbin, 2013), allowing for higher peak maxima than a typical GC separation (Baidoo & Teixeira Benites, 2019). On the other hand, LC-MS has excellent sensitivity for analyzing non-volatile substances (Navarro-Reig et al., 2017). Furthermore, regardless of the type of the LC column used, LC separation is prone to retention time variations when the material constitution is complex. LC-MS analysis is typically performed using ESI (electrospray ionization) and HRMS (high-resolution MS), which offer very precise mass estimations of the antecedent and fragment ions for metabolic detection, moreover, untargeted metabolomics is well performed with HRMS (Gross, 2011b, a).

CE-MS (capillary electrophoresis-based MS), in which components of investigated samples are differentiated based on mass-by-charge ratios, is a viable alternative to GC-MS and LC-MS and has promising potential for metabolomics (Aliferis & Chrysayi-Tokousbalides, 2011; Kostopoulou et al., 2020). Furthermore, developing two prominent MS sensors in metabolomics technologies, FT-ICR/MS (Fourier transform ion cyclotron resonance/MS) and Orbitrap MS, opened up new frontiers for metabolomics-based global physiological network research. The FT-ICR/MS is a powerful MS detector with great resolving power, high speed and precision. High-throughput metabolomics studies have effectively used such properties (Taylor et al., 2009).

4.2.6 INTEGRATED NMR AND MS-BASED METABOLOMICS

It is quite obvious that the development of metabolomics technologies and the advancement in artificial intelligence-based metabolomics software have eased and enhanced our ability to analyze complex compounds within no time. Despite the progress in metabolomics research, there are many metabolic compounds, and no single tool or software alone can analyze such a bulk of metabolic data. So, the best way is to integrate different metabolomic approaches to get the best result. Recent metabolomics studies have taken advantage of the complementarity among NMR and MS and between different MS devices, and it has proven to be very advantageous for metabolomics research (Aliferis & Jabaji, 2010; Aliferis & Chrysayi-Tokousbalides, 2011). In the field of metabolomics, MS and NMR-based metabolomics, as well as other combination techniques, have proven critical, facilitating the investigation of the metabolome of a wide range of species and the introduction of metabolomics to a variety of sectors (Huebschmann, 2015; Larive et al., 2015; Matich et al., 2019; Misra, 2019). Table 4.1 shows various metabolomics studies detecting different environmental stressors in different model organisms.

4.3 ENVIRONMENTAL METABOLOMICS

Environmental metabolomics is the application of metabolomics methods in understanding an organism's metabolic responses to environmental stimuli, including both biotic and abiotic stressors (Kovacevic & Simpson, 2020; Zhang et al., 2021). Life forms and the living environment are two critical aspects of environmental metabolomics research. Changes in metabolites (which leads to changes in metabolism) can give information about an organism's health and physiological condition, as well as assist how stressors impact the organism's wellness and hemodynamic state (Aliferis & Chrysayi-Tokousbalides, 2011; Matich et al., 2019; Jeevanandam & Osborne, 2021). The advancements in my metabolomic research have made it possible to analyze the impact of diverse environmental stressors on the physiology and biochemistry of organisms due to their high specificity and reproducibility (Zhang et al., 2018). After exposure to various environmental stimuli, such as organic pollutants (Xenobiotics such as heavy metals, pesticides, pharmaceuticals, and daily care products, and different persistent organic pollutants) and other naturally occurring components, environment metabolomics has been applied to a diverse array of biological entities (Raju & Bidlan, 2017;

TABLE 4.1 Metabolomics Methods for Detection of Environmental Stressors

Environmental Stressor	Detection Method	Concentration Detected	Model Organism/ Microbes	References
Zinc (heavy metal)	GC-TOF-MS	1,000 µg/L	*Tabellaria flocculosa* (diatom)	Gonçalves et al. (2018)
Copper (heavy metal)	LC-MS	3–6 mM	*Saccharomyces cerevisiae* (yeast)	Farrés et al. (2016)
Silver nanoparticles	GC-Q-TOF-MS GC-MS	10–100 µg/L 100 mg/kg soil	*Scenedesmus obliquus* (algae) Soil microbes	Wang et al. (2019b); Zhang et al. (2020)
TiO_2 & ZnO (nanoparticles)	GC-TOF-MS	10 & 0.2 mg/L, respectively	*Escherichia coli*	Pathakoti et al. (2020)
Chlorpyrifos (pesticide)	GC-MS	50 mg/L	*Mesorhizobium* sp.	Jabeen et al. (2015)
2,4-D (pesticide)	GC-MS	10 mM	*Escherichia coli*	Bhat et al. (2015)
Imidacloprid (pesticide)	NMR	0.25–0.05 mg/kg in soil test	*Eisenia fetida* (Earthworms)	Dani et al. (2019)
Acetamiprid	GC-MS	1/10 LC_{50}	*Brachydanio rerio* (Zebrafish)	Zhang & Zhao (2017)
Chlorpyrifos (pesticide)	HPLC	300 mg/L	*Bacillus pumilus*	Dar et al. (2019)
Sulfamethoxazole (pharmaceutical drug)	HPLC-HRMS	10 µg/L	*Mytilus galloprovincialis* (Mussels)	Serra-Compte et al. (2019)
Bisphenol A	NMR	1 mg/L	*Chironomus riparius* (midges)	Lee et al. (2018)
Mercury (heavy metal)	1H-13C-NMR	0.77 mg/kg	*Eisenia fetida* (earthworms)	Tang et al. (2018)
POPs (persistent organic pollutants)	GC-MS	2.0 mg/kg/d	Rats	Wang et al. (2018)
MoS_2	GC-MS	1,000 µg/mL	*Escherichia coli*	Wu et al. (2016)

Tomonaga et al., 2018; Roszkowska et al., 2018; Creydt et al., 2019; Xu et al., 2019; Naz et al., 2022).

One of several aims of environmental metabolomics investigation is to use metabolomics tools on keystone species that play an essential role in trophic levels and nutrient cycles for ecological biomonitoring and threat assessment (Bahamonde et al., 2016; Kovacevic & Simpson, 2020). Design of the study, exposure (collection of the environmental sample), sample preparation (control and perturbed environmental sample), metabolite retrieval, collection of data, analysis of data (using different tools such as

NMR, MS or both), and finally a biological exegesis of the analyzed data using appropriate bioinformatic tools are all part of the environmental metabolomics workflow (Sivaram et al., 2019). The implementation of SOPs (standard operating procedures) in metabolomics remains a concern since all metabolites cannot yet be determined using a single extraction process. One of the most important aspects of designing a robust metabolomics process is optimizing the extraction technique (Matich et al., 2019).

4.3.1 MAJOR ENVIRONMENTAL STRESSORS AND METABOLOMICS

Major environmental stressors are artificial by-products such as xenobiotics, including pesticides, heavy metals, persistent organic pollutants, pharmaceutical wastes, daily care products, dyes, oil spills, etc., which humans intentionally or unintentionally released into the environment. These compounds are toxic and contaminate the various components of the ecosystem such as air, water, land, and soil, etc., and also affect humans and other organisms such as cattle, fishes, birds, and microbes residing in the soil (Raj et al., 2021; Malla et al., 2022). Here we will briefly discuss the role of different pollutants/stressors in contaminating the environment the role of metabolomics in dealing with such stressors.

4.3.2 PESTICIDES

Pesticides are chemical compounds used to eradicate undesirable pests such as beetles, bugs, aphids, flies, rodents, nematodes, and a variety of fungal agents and undesired plants (weeds), to preserve plant health and boost agricultural output on the constantly dwindling land area (Lamilla et al., 2021). According to FAO data, India ranks second among Asian countries in pesticide use and fourth overall globally, behind China, the United States, and Japan. In the year 2020–2021, India consumed around 62,193 metric tons of pesticide, a quantity that continues to rise year after year (Chandran et al., 2020; FAOSTAT, 2021). In a study, organochlorine pesticides such as endosulfan, aldrin, endrin, dieldrin, endosulfan, and heptachlor were found in surface water, sediment, fish, and other aquatic biodiversity done by a group of experts in the Thamirabarani river basin in southern India (Raj et al., 2021). Organochlorines were identified by GC-MS after extraction using the QuEChERS technique (Arisekar et al., 2018).

Furthermore, the ongoing release of pesticides into the environment can have protracted or chronic consequences that are unknown. Plants were exposed to several pesticides in the lab, and the inadvertent impact on the plant's metabolome has been assessed by several researchers using metabolomic tools such as GC-MS and NMR (Varjani & Upasani, 2016; Campanale et al., 2021). GC-LRMS has been used to assess the metabolite alterations in rice following exposure to the pesticide diazinon. After treatment, alterations in metabolites involved in the production and metabolism of polyphenols, glucose, peptides, organic acids, and changes related to antioxidant defense, impairment with energy homeostasis, and oxidative stress, were detected (Mahdavi et al., 2015). Similar metabolic investigations were carried out using ^{1}H HRMAS-NMR-based metabolomics to explore the impact of chlordecone and lindane on maize roots, and various biochemical changes were observed, as well as changes in TCA (tri-carboxylic acid) cycle metabolites such as succinate and fumarate, suggesting a possible decrease in root respiration rate (Blondel et al., 2016).

4.3.3 NANOMATERIALS

Fabricated nanomaterials have been used in a broad range of applications as nanotechnology advances, raising concerns about their possible toxicity and off-target impacts on the environment (Soni et al., 2015). Nanoparticles are particles with a diameter ranging between 1 and 100 nm in at least one dimension and have been used in environmental remediation, electronics, healthcare, and medicine, energy applications, and a wide range of consumer items. Nanomaterials are projected to be discharged into the environment due to their extensive range of uses in many fields (Home | National Nanotechnology Initiative, 2022; Keller et al., 2013). Metabolomics has also been utilized to explore the metabolic alterations that nanomaterials may cause in non-target species, as well as the probable mode of action and toxicity of various nanomaterials on flora, fauna, and aquatic life. The metabolomics profiles of *Daphnia similis* exposed with silver nanoparticles (AgNPs) and $AgNO_3$ (Ag^+) were compared, and it was discovered that Ag^+ is the major culprit for acute intervention with energy metabolism and increased lipid peroxidation.

Arachidonic acid, a critical fatty acid in regulating sex ratios, is reduced in *Daphnia similis* after prolonged silver nanoparticles exposure (Li et al., 2015; Zhang et al., 2017). Researchers discovered significant changes in the metabolome of zebrafish (*Danio rerio*) on treatment with graphene-oxide based nanoparticles, particularly in the production of fatty acids and the tricarboxylic acid cycle (Sun et al., 2019). Using HRMS, the metabolite alterations in *Phaseolus vulgaris* were investigated following a 15-day soil or spray treatment of CeO_2. As per the findings, the spray application of CeO_2 had a stronger impact than the soil treatment (Matich et al., 2019). A group of researchers used metabolomics to examine the biochemical changes in spinach leaves after foliar spraying $Cu(OH)_2$ nano-insecticides for seven days. Metabolomics revealed that tocopherol, ascorbic acid, phenolic compounds, ferulic acid, and other metabolites were downregulated (Zhao et al., 2017).

4.3.4 PHARMACEUTICALS AND DAILY CARE PRODUCTS

Pharmaceuticals and hygiene products (PHPs) are routinely utilized in large quantities worldwide. They are continuously released into home and industrial sewer systems, and traditional wastewater treatment plants can only remove a small percentage of them. Because of their quasi-persistence, universality, and bioactivity in the environment, as well as their unanticipated negative effects on the ecosystem, PHPs have sparked widespread alarm as new pollutants (Abdul Mottaleb et al., 2015; Kim et al., 2019). The quantities of various medications and daily care items in drinking water, effluent, sea life, sewage, and plants have been determined by different researchers (Bartrons & Peñuelas, 2017, 2019). Due to the sheer variety of medications and daily care items that are discharged into the environment, there is an urgent need to investigate their harmful effects as well as their physiological interactions with non-target species. Significant alterations in the oxidative metabolism of fish and invertebrates have been observed in metabolomics investigations. Sulfamethazine, an antibiotic often used to treat infections caused by bacteria, has been shown to disrupt amino acid profiles in the *Oryzias melastigma* (Liu et al., 2018).

Even though many pharmacological metabolomics investigations have centered on fish and crustaceans, several researchers have begun to adapt metabolomics procedures to plant systems. The biotransformation products of the NSAID ibuprofen in *Vigna unguiculata* seeds, shoots, and roots, were identified using LC-HRMS in research. The parent molecule was identified, as well as 46 metabolites, including hydroxy derivatives and copolymers with hexoses and amino acids (Picó et al., 2018). Another investigation used

2D GC-HRMS to investigate the morphological and physiological impacts, as well as metabolic alterations, in lettuce *Lactuca sativa* subjected to 11 pollutants of high threat and concern (Matich et al., 2019). Based on the results of the above studies, it appears that metabolomics has presented a substitute for understanding the effects of pharmaceutical drugs and hygiene products in plants that have not been illustrated at sublethal concentrations via other ecotoxicological evaluations.

4.3.5 *HEAVY METALS*

Even though metals are essential for various biological processes such as energy flow and nutrient cycling, their excessive presence in the environment has the potential to harm a variety of ecosystem components. Due to anthropogenic activities, a wide variety of heavy metals, including arsenic, cadmium, chromium, copper, mercury, and others, are released into the environment, influencing a varied set of species in the ecosystem, according to different metabolomics research (Zhang et al., 2021). In different agro-ecological indices, heavy metal contamination has resulted in a significant deterioration in soil and water quality (Harrison et al., 2007). As per metabolomics findings, heavy metals cause alterations in metabolism linked to osmotic stress and physiology. The protective mechanism adapted by an organism under the stress of heavy metals can be deciphered by investigating the metabolites produced by an organism in its response (Nguyen et al., 2018). Interaction with heavy metals for a longer duration affects plants' metabolism by generating ROS (reactive oxygen species), which disrupts various biochemical pathways in plants and can be studied using metabolomics (Feng et al., 2021).

4.4 MICROBIAL METABOLOMICS

Microbes are ubiquitous in the environment and are used as pollution indicators as they suffer cytotoxic and genotoxic alterations when exposed to anthropogenic pollutants such as heavy metals, pesticides, nanoparticles, and other contaminants. Due to continuous interaction with such toxicants, microbes undergo mutational changes, thus developing resistance against these chemicals. These microbes feed on these xenobiotics as their carbon source and break them into non-toxic/less toxic by-products (Raj et al., 2021). Several metabolomics techniques such as GC-MS, LC-MS, GC-TOF-MS, etc., have revealed the metabolic changes occurring in the microbes and the biochemical changes occurring in the toxicant molecule/compound. Microorganisms are the most economical and sustainable solution to get rid of contaminated soil and water bodies. Recent developments in omics technologies have allowed researchers to access the particular gene/enzyme involved in degrading particular xenobiotics using the metagenomics/metabolomics approach (Narayanasamy et al., 2015; Aguiar-Pulido et al., 2016; Bilal et al., 2018). The majority of xenobiotics cause increased oxidative stress in the microbes exposed to them, thus producing/enhancing the level of ROS. A group of researchers has reported similar results in *E. coli.* when *E. coli* was exposed to ZnO and TiO_2 type nanoparticles, a higher ROS level was recorded using metabolomics techniques such as GC-TOF-MS. Very high fluctuations have been recorded in the functioning of several amino acids such as glycine, arginine, serine, proline, and threonine.

On further integration of the MS-based approach with NMR, the metabolites & proteins responsible for the elevated level of ROS in *E. coli* have been identified (Pathakoti et al., 2013, 2020; Planchon et al., 2017). Various soil microbes have to potential to degrade different types of soil contaminants such as pesticides, heavy metals, and hydrocarbons and have been reported by several researchers using the metabolomics approach (Yair et al., 2008; Khan, 2013; Jeffries et al., 2018).

4.5 WEB RESOURCES, DATABASES, AND SOFTWARE FOR METABOLOMICS RESEARCH

During the last several years, a considerable variety of freeware web-based resources and tools for metabolomics data pre-processing, analysis, imaging, and scientific interpretation have emerged, opening up new frontiers in metabolomics (Misra, 2019). The microorganisms can synthesize and bioconvert hazardous chemicals into harmless ones. There is a variety of software available for detecting chemicals, both free and commercial software. This program identifies the chemicals and may also assist in developing a suitable route for their degradation (Jeevanandam & Osborne, 2021). Table 4.2 contains a comprehensive list of such sources. Metabolomics research generates a huge number of data that demands time and effort to analyze biologically, as well as high-throughput techniques to correlate with biological processes and metabolic pathways of the natural systems under investigation. A few software solutions that permit the reliable integration of metabolomics datasets into metabolic networks are accessible, in contrast to

TABLE 4.2 List of Web Sources/Software for Metabolomic Data Analysis

Sl. No.	Software/Web Source	Applications	Link	References
1.	MS-FINDER	Characterization of compounds for GC-MS and MS-MS spectral analysis. Annotate unknowns using publicly available spectrum datasets like MassBank and LipidBlast.	http://prime.psc.riken.jp/compms/msfinder/main.html	Blaženović et al. (2018)
2.	COMSPARI	COMSPARI is a useful tool for comparing "matching" datasets.	https://bioinformaticshome.com/tools/proteomics/descriptions/COMSPARI.html	Kusonmano et al. (2016)
3.	BioSpyder	Data about biomolecules is tabulated and summarized using this tool.	https://www.biospyder.com/	Knox et al. (2007)
4.	BioTransformer	BioTransformer is a publicly downloadable software program for *in silico* metabolic prediction and molecule identification that is accurate, quick, and thorough.	http://biotransformer.ca/	Djoumbou-Feunang et al. (2019)
5.	COLMAR	To learn about the experimental NMR chemical shifts of well-known metabolites.	http://spin.ccic.ohio-state.edu/index.php/colmar	Wang et al. (2019a)
6.	MET-IDEA	MET-IDEA is a new data extraction tool for mass spectrometry-based metabolomics that quickly produces accurate and complete datasets.	https://met-idea.software.informer.com/	Lei et al. (2012)
7.	Open MS	An open-source C++ library for managing LC-MS data that includes a variety of algorithms for proteomics and metabolomics.	https://www.openms.de/	Rurik et al. (2020)
8.	MeltDB	A web-based software platform for analyzing and annotating metabolomics experiment datasets.	https://meltdb.cebitec.uni-bielefeld.de/cgi-bin/login.cgi	Neuweger et al. (2008)
9.	PRIMe	PRIMe (a RIKEN'S SUPPORTING PLATFORM) Metabolomics is a Web-based service that uses metabolomics and transcriptomics as systems to help people better comprehend life.	http://prime.psc.riken.jp/	Sakurai et al. (2013)

TABLE 4.2 *(Continued)*

Sl. No.	Software/Web Source	Applications	Link	References
10.	Omic Space	Integration and analysis of genomic phenomic connections based on coordinates.	http://omicspace.riken.jp/gps/	Toyoda et al. (2004)
11.	T3DB	It's a one-of-a-kind bioinformatics resource that combines thorough toxin data with extensive toxin target data. There are now 3,678 toxins in the database, which are related to 2,073 toxin target records and are described by 41,602 synonyms, including pollutants, pesticides, pharmaceuticals, and dietary poisons.	http://www.t3db.ca/	Uppal et al. (2017)
12.	METLIN exposome database	MS data in both positive and negative ionization modes, at numerous collision energy, is now available as searchable composite spectra, combining MS data from several collision energies into a single spectrum.	https://metlin.scripps.edu/landing_page.php?pgcontent=mainPage	Warth et al. (2017)
13.	Metabo Analyst	MetaboAnalyst is a comprehensive platform dedicated to metabolomics data analysis via a user-friendly, web-based interface.	https://www.metaboanalyst.ca/	Chong & Xia (2020)
14.	MZmine	MZmine is an open-source program for processing mass-spectrometry data, with an emphasis on LC-MS data.	http://mzmine.github.io/	Du et al. (2020)
15.	ChemSpider	ChemSpider is a free chemical structure database that allows users to explore over 100 million structures from hundreds of data sources using text and structure searches.	http://www.chemspider.com/	Uppal et al. (2017)
16.	SetupX	SetupX is a metabolomic LIMS platform that supports recording important GC-MS metabolomic information.	https://fiehnlab.ucdavis.edu/projects/binbase-setup	Scholz & Fiehn (2007)
17.	Met House	A data warehouse and repository for mass spectrometry metabolomics data that is vendor-independent.	https://msbi.ipb-halle.de/msbi/metware/methouse	Gaida & Neumann (2017)

TABLE 4.2 *(Continued)*

Sl. No.	Software/Web Source	Applications	Link	References
18.	MetaboMiner	It's a quick, non-destructive method for identifying and/or quantifying a large range of chemicals without having to separate them beforehand.	http://wishart.biology.ualberta.ca/metabominer/	Xia et al. (2008)
19.	PathVisio	It's a free open-source pathway designing and analysis program that lets you design, alter, and analyze biological pathways.	https://pathvisio.org/	Kutmon et al. (2015)

software development for database pre-processing and investigations. The program Cytoscape enables various high-throughput metabolomics techniques, which includes plug-ins like Metascape and MetaNetter (Aliferis & Chrysayi-Tokousbalides, 2011). The introduction of innovative, powerful, and advanced software that integrates numerous species-specific datasets is predicted to accelerate "omics" (particularly metabolomics) investigation and assist in their evolution as a strong global metabolomics tool for studying natural systems.

4.6 CONCLUSION AND FUTURE PERSPECTIVE

The global demand for increased food production and industrialization has contaminated the environmental components such as air, water, and other parts of the ecosystem, such as humans and other organisms. Moreover, the extensive use of chemical fertilizers and pesticides and accumulation of heavy metals in soil has adulterated the crop plants due to its uptake, and it further reaches the food chain, thus affecting the health of humans and other species. However, the emergence of omics technologies (especially metabolomics) has solved the concern to deal with such pollutants to a greater extent. The development in the field of metabolomics instrumentation such as GC-MS, NMR, and HPLC sophisticated instruments, has helped researchers decipher the fate and action of environmental stressors on different organisms and plants. Apart from the advancement in metabolomics, artificial intelligence and bioinformatics have developed at a greater pace. Its integration with metabolomics has enhanced the pace of research and helps predict the pathway of certain chemical compounds and their interaction with different components of the environment. Nonetheless, the absence of standardized analytical techniques for metabolomics and an inadequate understanding of plant metabolic pathways are drawbacks that must be solved to standardize metabolomics. Besides that, the interconnection of various analysis tools for multi-complex sample studies and the interconnection of metabolomics with other "omics" strategies in the context of a high-dimensional biological approach is expected to yield novel clues into the functioning of biological pathways.

Moreover, there is a need for a technological revolution in the field of metabolomics as most of the tools in action are either too expensive or require high maintenance cost, which needs to be resolved, and a more user-friendly device needs to be discovered. Apart from this, there is also a need for software development that is freely accessible and solves most of the purposes on a single platform.

ACKNOWLEDGMENTS

Aman Raj would like to acknowledge Indian Council of Medical Research, New Delhi, India for providing financial support through ICMR-SRF fellowship.

KEYWORDS

- **agrochemicals**
- **correlation spectroscopy**
- **GC-MS**
- **metabolomics**
- **nuclear magnetic resonance**
- **total correlation spectroscopy**
- **xenobiotics**

REFERENCES

Abdul, M. M., Abdul, M. M., Md Abdul, W., et al., (2015). *Emerging Micro-pollutants Pharmaceuticals and Personal Care Products (PPCPs) Contamination Concerns in Aquatic Organisms-LC/MS and GC/MS Analysis* (Vol. 1198, pp. 43–74). ACS Publ. https://doi.org/10.1021/bk-2015-1198.ch003.

Aga, D. S. (Ed.). (2007). Fate of pharmaceuticals in the environment and in water treatment systems. CRC Press. https://doi.org/10.1201/9781420052336.

Aguiar-Pulido, V., Huang, W., Suarez-Ulloa, V., et al., (2016). Metagenomics, metatranscriptomics, and metabolomics approaches for microbiome analysis. *Evol. Bioinforma., 12*, 5–16. https://doi.org/10.4137/EBO.S36436.

Aliferis, K. A., & Chrysayi-Tokousbalides, M., (2011). Metabolomics in pesticide research and development: Review and future perspectives. *Metabolomics, 7*, 35–53. https://doi.org/10.1007/S11306-010-0231-X.

Aliferis, K. A., & Jabaji, S., (2010). 1H NMR and GC-MS metabolic fingerprinting of developmental stages of *Rhizoctonia solani* sclerotia. *Metabolomics, 6*, 96–108. https://doi.org/10.1007/S11306-009-0180-4.

Al-Zubaidi, L. A., Alnuaimi, M. T., Aljanabi, Z. Z., & Adel, M. M., (2021). Biodegradation of chlorpyrifos pesticide using silver bio-nanoparticles *Bacillus thuringiensis israelensis* extracts. *IOP Conf. Ser. Earth Environ. Sci., 779*, 012113. https://doi.org/10.1088/1755-1315/779/1/012113.

Aretz, I., & Meierhofer, D., (2016). Advantages and pitfalls of mass spectrometry-based metabolome profiling in systems biology. *Int. J. Mol. Sci., 17*. https://doi.org/10.3390/IJMS17050632.

Arisekar, U., Shakila, R. J., Jeyasekaran, G., Shalini, R., Kumar, P., Malani, A. H., & Rani, V. (2019). Accumulation of organochlorine and pyrethroid pesticide residues in fish, water, and sediments in the Thamirabarani river system of southern peninsular India. *Environmental Nanotechnology, Monitoring and Management, 11,* 100194. https://doi.org/10.1016/j.enmm.2018.11.003.

Bahamonde, P. A., Feswick, A., Isaacs, M. A., et al., (2016). Defining the role of omics in assessing ecosystem health: Perspectives from the Canadian environmental monitoring program. *Environ. Toxicol. Chem., 35*, 20–35. https://doi.org/10.1002/ETC.3218.

Baidoo, E. E. K., & Teixeira, B. V., (2019). Mass spectrometry-based microbial metabolomics: Techniques, analysis, and applications. *Methods Mol. Biol., 1859*, 11–69. https://doi.org/10.1007/978-1-4939-8757-3_2.

Bartrons, M., & Peñuelas, J., (2017). Pharmaceuticals and personal-care products in plants. *Trends Plant Sci., 22*, 194–203. https://doi.org/10.1016/J.TPLANTS.2016.12.010.

Bhat, S. V., Booth, S. C., Vantomme, E. A. N., et al., (2015). Oxidative stress and metabolic perturbations in *Escherichia coli* exposed to sublethal levels of 2,4-dichlorophenoxyacetic acid. *Chemosphere, 135*, 453–461. https://doi.org/10.1016/J.CHEMOSPHERE.2014.12.035.

Bilal, T., Malik, B., & Hakeem, K. R., (2018). Metagenomic analysis of uncultured microorganisms and their enzymatic attributes. *J. Microbiol. Methods, 155*, 65–69. https://doi.org/10.1016/J.MIMET.2018.11.014.

Blasco, H., Corcia, P., Moreau, C., et al., (2010). 1H-NMR-based metabolomic profiling of CSF in early amyotrophic lateral sclerosis. *PLoS One, 5*. https://doi.org/10.1371/JOURNAL.PONE.0013223.

Blaženović, I., Kind, T., Ji, J., & Fiehn, O., (2018). Software tools and approaches for compound identification of LC-MS/MS data in metabolomics. *Metab., 8,* 31. https://doi.org/10.3390/METABO8020031.

Blondel, C., Khelalfa, F., Reynaud, S., et al., (2016). Effect of organochlorine pesticides exposure on the maize root metabolome assessed using high-resolution magic-angle spinning (1)H NMR spectroscopy. *Environ. Pollut., 214*, 539–548. https://doi.org/10.1016/J.ENVPOL.2016.04.057.

Boesten, J. J. T. I., (2016). Proposal for field-based definition of soil bound pesticide residues. *Sci. Total Environ.*

Burton, I. W., Quilliam, M. A., & Walter, J. A., (2005). Quantitative 1H NMR with external standards: Use in preparation of calibration solutions for algal toxins and other natural products. *Anal. Chem., 77*, 3123–3131. https://doi.org/10.1021/AC048385H/SUPPL_FILE/AC048385HSI20050128_044400.PDF.

Campanale, C., Massarelli, C., Losacco, D., et al., (2021). The monitoring of pesticides in water matrices and the analytical criticalities: A review. *TrAC – Trends Anal. Chem., 144*, 116423. https://doi.org/10.1016/j.trac.2021.116423.

Cañueto, D., Gómez, J., Salek, R. M., et al., (2018). rDolphin: A GUI R package for proficient automatic profiling of 1D 1 H-NMR spectra of study datasets. *Metabolomics, 14*. https://doi.org/10.1007/S11306-018-1319-Y.

Chandran, H., Meena, M., Barupal, T., & Sharma, K., (2020). Plant tissue culture as a perpetual source for production of industrially important bioactive compounds. *Biotechnol. Reports, 26*. https://doi.org/10.1016/J.BTRE.2020.E00450.

Chong, J., & Xia, J., (2020). Using MetaboAnalyst 4.0 for metabolomics data analysis, interpretation, and integration with other omics data. *Methods Mol Biol, 2104*, 337–360. https://doi.org/10.1007/978-1-0716-0239-3_17.

Cloarec, O., Dumas, M. E., Craig, A., et al., (2005). Statistical total correlation spectroscopy: An exploratory approach for latent biomarker identification from metabolic 1H NMR data sets. *Anal. Chem., 77*, 1282–1289. https://doi.org/10.1021/AC048630X.

Creydt, M., Vuralhan-Eckert, J., Fromm, J., & Fischer, M., (2019). Effects of elevated CO_2 concentration on leaves and berries of black elder (*Sambucus nigra*) using UHPLC-ESI-QTOF-MS/MS and gas exchange measurements. *J. Plant Physiol., 234, 235*, 71–79. https://doi.org/10.1016/J.JPLPH.2019.01.005.

Crook, A. A., & Powers, R., (2020). Quantitative NMR-based biomedical metabolomics: Current status, and applications. *Molecules, 25*. https://doi.org/10.3390/MOLECULES25215128.

Dani, V. D., Lankadurai, B. P., Nagato, E. G., et al., (2019). Comparison of metabolomic responses of earthworms to sub-lethal imidacloprid exposure in contact and soil tests. *Environ. Sci. Pollut. Res., 26*, 18846–18855. https://doi.org/10.1007/S11356-019-05302-Y/FIGURES/4.

Dar, M. A., Kaushik, G., & Villarreal-Chiu, J. F., (2019). Pollution status and bioremediation of chlorpyrifos in environmental matrices by the application of bacterial communities: A review. *J. Environ. Manage, 239,* 124–136. https://doi.org/10.1016/J.JENVMAN.2019.03.048.

Dettmer, K., Aronov, P. A., & Hammock, B. D., (2007). Mass spectrometry-based metabolomics. *Mass Spectrom. Rev., 26*, 51–78. https://doi.org/10.1002/MAS.20108.

Djoumbou-Feunang, Y., Fiamoncini, J., Gil-de-la-Fuente, A., et al., (2019). BioTransformer: A comprehensive computational tool for small molecule metabolism prediction and metabolite identification. *J. Cheminform., 11*. https://doi.org/10.1186/S13321-018-0324-5.

Du, X., Smirnov, A., Pluskal, T., et al., (2020). Metabolomics data preprocessing using ADAP and MZmine 2. *Methods Mol. Biol., 2104*, 25–48. https://doi.org/10.1007/978-1-0716-0239-3_3.

Dubey, A., Malla, M. A., Khan, F., et al., (2019). Soil microbiome: A key player for conservation of soil health under changing climate. *Biodivers. Conserv., 28*, 2405–2429. https://doi.org/10.1007/s10531-019-01760-5.

Dunn, W. B., & Ellis, D. I., (2005). Metabolomics: Current analytical platforms and methodologies. *TrAC Trends Anal. Chem., 24*, 285–294. https://doi.org/10.1016/J.TRAC.2004.11.021.

Emwas, A. H. M., (2015). The strengths and weaknesses of NMR spectroscopy and mass spectrometry with particular focus on metabolomics research. *Methods Mol. Biol, 1277*, 161–193. https://doi.org/10.1007/978-1-4939-2377-9_13.

Emwas, A. H., Roy, R., McKay, R. T., et al., (2019). NMR spectroscopy for metabolomics research. *Metab., 9*, 123. https://doi.org/10.3390/METABO9070123.

FAOSTAT (2021). *FAOSTAT.* Available online at: https://www.fao.org/faostat/en/#home (accessed on 15 November 2023).

Farrés, M., Piña, B., & Tauler, R., (2016). LC-MS based metabolomics and chemometrics study of the toxic effects of copper on *Saccharomyces cerevisiae*. *Metallomics, 8*, 790–798. https://doi.org/10.1039/C6MT00021E.

Feng, Z., Ji, S., Ping, J., & Cui, D., (2021). Recent advances in metabolomics for studying heavy metal stress in plants. *TrAC Trends Anal. Chem., 143*, 116402. https://doi.org/10.1016/J.TRAC.2021.116402.

Gaida, A., & Neumann, S., (2017). MetHouse: Raw and preprocessed mass spectrometry data. *J. Integr. Bioinform., 4*, 107–114. https://doi.org/10.1515/JIB-2007-56/PDF.

Gonçalves, S., Kahlert, M., Almeida, S. F. P., & Figueira, E., (2018). A freshwater diatom challenged by Zn: Biochemical, physiological and metabolomic responses of *Tabellaria flocculosa* (Roth) Kützing. *Environ. Pollut., 238*, 959–971. https://doi.org/10.1016/J.ENVPOL.2018.01.111.

Gross, J. H., (2011a). Practical aspects of electron ionization. *Mass Spectrom.*, 223–248. https://doi.org/10.1007/978-3-642-10711-5_5.

Gross, J. H., (2011b). *Mass Spectrometry: A Textbook* (2nd edn., pp. 1–753). https://doi.org/10.1007/978-3-642-10711-5.

Hao, J., Astle, W., De iorio, M., & Ebbels, T. M. D., (2012). BATMAN—An R package for the automated quantification of metabolites from nuclear magnetic resonance spectra using a Bayesian model. *Bioinformatics, 28*, 2088–2090. https://doi.org/10.1093/BIOINFORMATICS/BTS308.

Harrison, J. J., Ceri, H., Yerly, J., et al., (2007). Metal ions may suppress or enhance cellular differentiation in *Candida albicans* and *Candida tropicalis* biofilms. *Appl. Environ. Microbiol., 73*, 4940–4949. https://doi.org/10.1128/AEM.02711-06/SUPPL_FILE/SUPPLEMENTARY_VIDEO_4.ZIP.

Heux, S., Bergès, C., Millard, P., et al., (2017). Recent advances in high-throughput 13 C-fluxomics. *Curr. Opin. Biotechnol., 43*, 104–109. https://doi.org/10.1016/J.COPBIO.2016.10.010.

Hoffmann, E. D. S. V., (2007). *Mass Spectrometry: Principles and Applications* (3rd edn., pp. 1–4). Edmond de Hoffmann, Vincent Stroobant.

Hoult, D. I., Busby, S. J. W., Gadian, D. G., et al., (1974). Observation of tissue metabolites using 31P nuclear magnetic resonance. *Nat., 252*, 285–287. https://doi.org/10.1038/252285a0.

Huebschmann, H. J., (2015). *Handbook of GC-MS: Fundamentals and Applications* (3rd edn., p. 880).

Jabeen, H., Iqbal, S., & Anwar, S., (2015). Biodegradation of chlorpyrifos and 3, 5, 6-trichloro-2-pyridinol by a novel rhizobial strain *Mesorhizobium* sp. HN3. *Water Environ. J., 29*, 151–160. https://doi.org/10.1111/WEJ.12081.

Jang, C., Chen, L., & Rabinowitz, J. D., (2018). Metabolomics and isotope tracing. *Cell, 173*, 822–837. https://doi.org/10.1016/J.CELL.2018.03.055.

Jeevanandam, V., & Osborne, J., (2021). Understanding the fundamentals of microbial remediation with emphasize on metabolomics. *Prep. Biochem. Biotechnol.*, 1–13. https://doi.org/10.1080/10826068.2021.1946694.

Jeffries, T. C., Rayu, S., Nielsen, U. N., et al., (2018). Metagenomic functional potential predicts degradation rates of a model organophosphorus xenobiotic in pesticide contaminated soils. *Front. Microbiol.* https://doi.org/10.3389/fmicb.2018.00147.

Keller, A. A., McFerran, S., Lazareva, A., & Suh, S., (2013). Global life cycle releases of engineered nanomaterials. *J. Nanoparticle. Res., 15*, 1–17. https://doi.org/10.1007/S11051-013-1692-4/FIGURES/14.

Khan, F., (2013). In silico approach for the bioremediation of toxic pollutants. *J. Pet. Environ. Biotechnol., 4*. https://doi.org/10.4172/2157-7463.1000161.

Kim, H. M., Long, N. P., Yoon, S. J., et al., (2019). Metabolomics and phenotype assessment reveal cellular toxicity of triclosan in *Caenorhabditis elegans*. *Chemosphere, 236*, 124306. https://doi.org/10.1016/J.CHEMOSPHERE.2019.07.037.

Kim, Y. M., & Heyman, H. M., (2018). Mass spectrometry-based metabolomics. *Methods Mol. Biol., 1775*, 107–118. https://doi.org/10.1007/978-1-4939-7804-5_10.

Knox, C., Shrivastava, S., Stothard, P., et al., (2007). Biospider: A web server for automating metabolome annotations. *Pacific Symp. Biocomput., 2007*, 145–156. https://doi.org/10.1142/9789812772435_0015.

Kono, H., (2013). 1H and 13C chemical shift assignment of the monomers that comprise carboxymethyl cellulose. *Carbohydr. Polym., 97*, 384–390. https://doi.org/10.1016/J.CARBPOL.2013.05.031.

Kostopoulou, S., Ntatsi, G., Arapis, G., & Aliferis, K. A., (2020). Assessment of the effects of metribuzin, glyphosate, and their mixtures on the metabolism of the model plant *Lemna minor* L. applying metabolomics. *Chemosphere, 239*, 124582. https://doi.org/10.1016/j.chemosphere.2019.124582.

Kovacevic, V., & Simpson, M. J., (2020). Fundamentals of environmental metabolomics. *Environ. Metabolomics,* 1–33. https://doi.org/10.1016/B978-0-12-818196-6.00001-7.

Kuehnbaum, N. L., & Britz-Mckibbin, P., (2013). New advances in separation science for metabolomics: Resolving chemical diversity in a post-genomic era. *Chem. Rev., 113*, 2437–2468. https://doi.org/10.1021/CR300484S.

Kumar, A., Dubey, A., Malla, M. A., & Dames, J., (2021). Pyrosequencing and phenotypic microarray to decipher bacterial community variation in *Sorghum bicolor* (L.) Moench rhizosphere. *Curr. Res. Microb. Sci.* https://doi.org/10.1016/j.crmicr.2021.100025.

Kusonmano, K., Vongsangnak, W., & Chumnanpuen, P., (2016). Informatics for metabolomics. *Adv. Exp. Med. Biol., 939*, 91–115. https://doi.org/10.1007/978-981-10-1503-8_5.

Kutmon, M., Van, I. M. P., Bohler, A., et al., (2015). PathVisio 3: An extendable pathway analysis toolbox. *PLoS Comput. Biol., 11*. https://doi.org/10.1371/JOURNAL.PCBI.1004085.

Lamilla, C., Schalchli, H., Briceño, G., et al., (2021). A pesticide biopurification system: A source of biosurfactant-producing bacteria with environmental biotechnology applications. *Agronomy, 11*, 624. https://doi.org/10.3390/agronomy11040624.

Larive, C. K., Barding, G. A., & Dinges, M. M., (2015). NMR spectroscopy for metabolomics and metabolic profiling. *Anal. Chem., 87*, 133–146. https://doi.org/10.1021/AC504075G/ASSET/IMAGES/AC504075G.SOCIAL.JPEG_V03.

Lee, S. W., Chatterjee, N., Im, J. E., et al., (2018). Integrated approach of eco-epigenetics and eco-metabolomics on the stress response of bisphenol-A exposure in the aquatic midge *Chironomus riparius*. *Ecotoxicol. Environ. Saf., 163*, 111–116. https://doi.org/10.1016/J.ECOENV.2018.06.084.

Lei, Z., Li, H., Chang, J., et al., (2012). MET-IDEA version 2.06; improved efficiency and additional functions for mass spectrometry-based metabolomics data processing. *Metabolomics, 8*, 105–110. https://doi.org/10.1007/S11306-012-0397-5.

Li, L. Z., Wu, H., Ji, C., et al., (2015). A metabolomic study on the responses of daphnia magna exposed to silver nitrate and coated silver nanoparticles. *Ecotoxicol. Environ. Saf., 119*, 66–73. https://doi.org/10.1016/J.ECOENV.2015.05.005.

Liu, Y., Wang, X., Li, Y., & Chen, X., (2018). Metabolomic analysis of short-term sulfamethazine exposure on marine medaka (*Oryzias melastigma*) by comprehensive two-dimensional gas chromatography-time-of-flight mass spectrometry. *Aquat. Toxicol., 198*, 269–275. https://doi.org/10.1016/J.AQUATOX.2018.03.006.

Lucio-Gutiérrez, J. R., Delgado-Montemayor, C., Coello-Bonilla, J., et al., (2019). Selective 1D-TOCSY and chemometrics to evaluate authenticity of *Turnera diffusa* and related botanical extracts. *Phytochem. Lett., 30*, 62–68. https://doi.org/10.1016/J.PHYTOL.2019.01.011.

Ludwig, C., Viant, M. R., Choi, Y. H., et al., (2010). Two-dimensional J-resolved NMR spectroscopy: Review of a key methodology in the metabolomics toolbox. *Phytochem. Anal., 21*, 22–32. https://doi.org/10.1002/PCA.1186.

Macura, S., Kumar, N. G., & Brown, L. R., (1983). Combined use of COSY and double quantum two-dimensional NMR spectroscopy for elucidation of spin systems in polymyxin B. *Biochem. Biophys. Res. Commun., 117*, 486–492. https://doi.org/10.1016/0006-291X(83)91226-3.

Mahdavi, V., Farimani, M. M., Fathi, F., & Ghassempour, A., (2015). A targeted metabolomics approach toward understanding metabolic variations in rice under pesticide stress. *Anal. Biochem., 478*, 65–72. https://doi.org/10.1016/J.AB.2015.02.021.

Malla, M. A., Dubey, A., Raj, A., et al., (2022). Emerging frontiers in microbe-mediated pesticide remediation: Unveiling role of omics and in silico approaches in engineered environment. *Environ. Pollut., 299*, 118851. https://doi.org/10.1016/j.envpol.2022.118851.

Malla, M. A., Dubey, A., Yadav, S., et al., (2018). Understanding and designing the strategies for the microbe-mediated remediation of environmental contaminants using omics approaches. *Front. Microbiol., 9*. https://doi.org/10.3389/fmicb.2018.01132.

Malla, M. A., Gupta, S., Dubey, A., et al., (2021). Contamination of groundwater resources by pesticides. *Contam. Water*, 99–107. https://doi.org/10.1016/B978-0-12-824058-8.00023-2.

Markley, J. L., Brüschweiler, R., Edison, A. S., et al., (2017). The future of NMR-based metabolomics. *Curr. Opin. Biotechnol., 43*, 34–40. https://doi.org/10.1016/J.COPBIO.2016.08.001.

Matich, E. K., Chavez, S. N. G., Aga, D. S., & Atilla-Gokcumen, G. E., (2019). Applications of metabolomics in assessing ecological effects of emerging contaminants and pollutants on plants. *J. Hazard Mater., 373*, 527–535. https://doi.org/10.1016/J.JHAZMAT.2019.02.084.

Melvin, S. D., Lanctôt, C. M., Doriean, N. J. C., et al., (2018). Untargeted NMR-based metabolomics for field-scale monitoring: Temporal reproducibility and biomarker discovery in mosquitofish (*Gambusia holbrooki*) from a metal(loid)-contaminated wetland. *Environ. Pollut., 243*, 1096–1105. https://doi.org/10.1016/J.ENVPOL.2018.09.071.

Misra, B. B., (2019). Metabolomics tools to study links between pollution and human health: An exposomics perspective. *Curr. Pollut. Reports, 5*, 93–111. https://doi.org/10.1007/S40726-019-00109-4.

Misra, B. B., Langefeld, C., Olivier, M., & Cox, L. A., (2019). Integrated omics: Tools, advances, and future approaches. *J. Mol. Endocrinol., 62*, R21–R45. https://doi.org/10.1530/JME-18-0055.

Narayanasamy, S., Muller, E. E. L., Sheik, A. R., & Wilmes, P., (2015). Integrated omics for the identification of key functionalities in biological wastewater treatment microbial communities. *Microb. Biotechnol., 8*, 363–368. https://doi.org/10.1111/1751-7915.12255.

National Nanotechnology Initiative, (2022). https://www.nano.gov/2022BudgetSupplement (accessed on 15 November 2023).

Navarro-Reig, M., Jaumot, J., Baglai, A., et al., (2017). Untargeted comprehensive two-dimensional liquid chromatography coupled with high-resolution mass spectrometry analysis of rice metabolome using multivariate curve resolution. *Anal. Chem., 89*, 7675–7683. https://doi.org/10.1021/ACS.ANALCHEM.7B01648/SUPPL_FILE/AC7B01648_SI_001.PDF.

Naz, M., Shah, T., Battaglia, M., et al., (2022). Insights into potential roles of plants as natural radioprotectants and amelioration of radiations induced harmful impacts on human health. *Manag. Plant Prod. Under Chang. Environ.*, 311–325. https://doi.org/10.1007/978-981-16-5059-8_12.

Neuweger, H., Albaum, S. P., Dondrup, M., Persicke, M., Watt, T., Niehaus, K., ... & Goesmann, A. (2008). MeltDB: a software platform for the analysis and integration of metabolomics experiment data. *Bioinformatics, 24*(23), 2726–2732. doi: 10.1093/bioinformatics/btn452.

Nguyen, T. V., Alfaro, A. C., Merien, F., et al., (2018). Copper-induced immunomodulation in mussel (*Perna canaliculus*) haemocytes. *Metallomics, 10*, 965–978. https://doi.org/10.1039/C8MT00092A.

Núñez, O., Gallart-Ayala, H., Martins, C. P. B., & Lucci, P., (2012). New trends in fast liquid chromatography for food and environmental analysis. *J. Chromatogr. A., 1228*, 298–323. https://doi.org/10.1016/J.CHROMA.2011.10.091.

Pathakoti, K., Manubolu, M., & Hwang, H. M., (2020). Mechanistic insights into TiO_2 and ZnO nanoparticle-induced metabolic changes in *Escherichia coli* under solar simulated light irradiation. *Water Air Soil Pollut., 231*. https://doi.org/10.1007/S11270-019-4388-2.

Pathakoti, K., Morrow, S., Han, C., et al., (2013). Photoinactivation of *Escherichia coli* by sulfur-doped and nitrogen-fluorine-codoped TiO_2 nanoparticles under solar simulated light and visible light irradiation. *Environ. Sci. Technol., 47*, 9988–9996. https://doi.org/10.1021/ES401010G/SUPPL_FILE/ES401010G_SI_001.PDF.

Pauling, L., Robinson, A. B., Teranishi, R., & Cary, P., (1971). Quantitative analysis of urine vapor and breath by gas-liquid partition chromatography. *Proc. Natl. Acad. Sci. U S A., 68*, 2374–2376. https://doi.org/10.1073/PNAS.68.10.2374.

Picó, Y., Alvarez-Ruiz, R., Wijaya, L., et al., (2018). Analysis of ibuprofen and its main metabolites in roots, shoots, and seeds of cowpea (*Vigna unguiculata* L. Walp) using liquid chromatography-quadrupole time-of-flight mass spectrometry: Uptake, metabolism, and translocation. *Anal. Bioanal. Chem., 410*, 1163–1176. https://doi.org/10.1007/S00216-017-0796-6.

Planchon, M., Leger, T., Spalla, O., et al., (2017). Metabolomic and proteomic investigations of impacts of titanium dioxide nanoparticles on *Escherichia coli*. *PLoS One, 12*, e0178437. https://doi.org/10.1371/JOURNAL.PONE.0178437.

Pöhö, P., Lipponen, K., Bespalov, M. M., et al., (2019). Comparison of liquid chromatography-mass spectrometry and direct infusion microchip electrospray ionization mass spectrometry in global metabolomics of cell samples. *Eur. J. Pharm. Sci., 138*, 104991. https://doi.org/10.1016/J.EJPS.2019.104991.

Prihandiani, A., Bella, D. R., Chairani, N. R., et al., (2021). *The Tsunami of Pesticide Use for Rice Production on Java and Its Consequences* (Vol. 22, pp. 276–297). https://doi.org/10.1080/14442213.2021.1942970.

Rahman, A. U., & Choudhary, M. I. (Eds.). (2015). Applications of NMR Spectroscopy: Volume 3 (Vol. 3). Elsevier.

Raj, A., Kumar, A., & Dames, J. F., (2021). Tapping the role of microbial biosurfactants in pesticide remediation: An eco-friendly approach for environmental sustainability. *Front. Microbiol., 12*. https://doi.org/10.3389/fmicb.2021.791723.

Raju, S. M., & Bidlan, R., (2017). Isolation, enrichment and metagenomic characterization of simultaneous DDT and lindane degrading microbial consortium. *J. Pure Appl. Microbiol., 11*, 1947–1952. https://doi.org/10.22207/JPAM.11.4.36.

Rohr, J. R., Farag, A. M., Cadotte, M. W., et al., (2016). Transforming ecosystems: When, where, and how to restore contaminated sites. *Integr, Environ. Assess. Manag., 12*, 273–283. https://doi.org/10.1002/ieam.1668.

Roszkowska, A., Yu, M., Bessonneau, V., et al., (2018). Metabolome profiling of fish muscle tissue exposed to benzo[a]pyrene using *in vivo* solid-phase microextraction. *Environ. Sci. Technol. Lett., 5*, 431–435. https://doi.org/10.1021/ACS.ESTLETT.8B00272/SUPPL_FILE/EZ8B00272_SI_001.PDF.

Rurik, M., Alka, O., Aicheler, F., & Kohlbacher, O., (2020). Metabolomics data processing using openMS. *Methods Mol. Biol., 2104*, 49–60. https://doi.org/10.1007/978-1-0716-0239-3_4.

Sakurai, T., Yamada, Y., Sawada, Y., Matsuda, F., Akiyama, K., Shinozaki, K., ... & Saito, K. (2013). PRIMe update: innovative content for plant metabolomics and integration of gene expression and metabolite accumulation. *Plant and Cell Physiology, 54*(2), e5–e5. doi: 10.1093/pcp/pcs184.

Sandusky, P., Appiah-Amponsah, E., & Raftery, D., (2011). Use of optimized 1D TOCSY NMR for improved quantitation and metabolomic analysis of biofluids. *J. Biomol, NMR., 49*, 281. https://doi.org/10.1007/S10858-011-9483-7.

Schätzlein, M. P., Becker, J., Schulze-Sünninghausen, D., et al., (2018). Rapid two-dimensional ALSOFAST-HSQC experiment for metabolomics and fluxomics studies: Application to a 13 C-enriched cancer cell model treated with gold nanoparticles. *Anal. Bioanal. Chem., 410*, 2793–2804. https://doi.org/10.1007/S00216-018-0961-6.

Scholz, M., & Fiehn, O., (2007). SetupX-a public study design database for metabolomic projects. *Pacific Symp. Biocomput., 2007,* 169–180. https://doi.org/10.1142/9789812772435_0017.

Sekiyama, Y., Chikayama, E., & Kikuchi, J., (2011). Evaluation of a semipolar solvent system as a step toward heteronuclear multidimensional NMR-based metabolomics for 13C-labeled bacteria, plants, and animals. *Anal. Chem., 83*, 719–726. https://doi.org/10.1021/AC102097U.

Serra-Compte, A., Álvarez-Muñoz, D., Solé, M., et al., (2019). Comprehensive study of sulfamethoxazole effects in marine mussels: Bioconcentration, enzymatic activities and metabolomics. *Environ. Res., 173*, 12–22. https://doi.org/10.1016/J.ENVRES.2019.03.021.

Sivaram, A. K., Subashchandrabose, S. R., Logeshwaran, P., et al., (2019). Metabolomics reveals defensive mechanisms adapted by maize on exposure to high molecular weight polycyclic aromatic hydrocarbons. *Chemosphere, 214*, 771–780. https://doi.org/10.1016/J.CHEMOSPHERE.2018.09.170.

Soni, D., Naoghare, P. K., Saravanadevi, S., & Pandey, R. A., (2015). Release, transport, and toxicity of engineered nanoparticles. *Rev. Environ. Contam. Toxicol., 234*, 1–47. https://doi.org/10.1007/978-3-319-10638-0_1.

Sun, J., Zhou, Q., & Hu, X., (2019). Integrating multi-omics and regular analyses identifies the molecular responses of zebrafish brains to graphene oxide: Perspectives in environmental criteria. *Ecotoxicol. Environ. Saf., 180*, 269–279. https://doi.org/10.1016/J.ECOENV.2019.05.011.

Tang, R., Ding, C., Dang, F., et al., (2018). NMR-based metabolic toxicity of low-level Hg exposure to earthworms. *Environ. Pollut., 239*, 428–437. https://doi.org/10.1016/J.ENVPOL.2018.04.027.

Tardivel, P. J. C., Canlet, C., Lefort, G., et al., (2017). ASICS: An automatic method for identification and quantification of metabolites in complex 1D 1H NMR spectra. *Metabolomics, 13*. Non Paginé. https://doi.org/10.1007/S11306-017-1244-5.

Taylor, N. S., Weber, R. J. M., Southam, A. D., et al., (2009). A new approach to toxicity testing in daphnia magna: Application of high throughput FT-ICR mass spectrometry metabolomics. *Metabolomics, 5*, 44–58. https://doi.org/10.1007/S11306-008-0133-3/TABLES/3.

Tomonaga, S., Okuyama, H., Tachibana, T., & Makino, R., (2018). Effects of high ambient temperature on plasma metabolomic profiles in chicks. *Anim. Sci. J., 89*, 448–455. https://doi.org/10.1111/ASJ.12951.

Toyoda, T., & Wada, A. (2004). Omic space: coordinate-based integration and analysis of genomic phenomic interactions. Bioinformatics, 20(11), 1759-1765. https://doi.org/10.1093/bioinformatics/bth165.

Uppal, K., Walker, D. I., & Jones, D. P., (2017). xMSannotator: An R package for network-based annotation of high-resolution metabolomics data. *Anal. Chem., 89*, 1063–1067. https://doi.org/10.1021/ACS.ANALCHEM.6B01214.

Varjani, S. J., & Upasani, V. N., (2016). Core flood study for enhanced oil recovery through ex-situ bioaugmentation with thermo- and halo-tolerant rhamnolipid produced by *Pseudomonas aeruginosa* NCIM 5514. *Bioresour. Technol., 220*, 175–182. https://doi.org/10.1016/j.biortech.2016.08.060.

Villas-Bôas, S. G., Mas, S., Åkesson, M., et al., (2005). Mass spectrometry in metabolome analysis. *Mass Spectrom. Rev., 24*, 613–646. https://doi.org/10.1002/MAS.20032.

Wang, C., Zhang, B., Timári, I., et al., (2019a). Accurate and efficient determination of unknown metabolites in metabolomics by NMR-based molecular motif identification. *Anal. Chem., 91*, 15686–15693. https://doi.org/10.1021/ACS.ANALCHEM.9B03849.

Wang, J., Li, C. L., Tu, B. J., et al., (2018). Integrated epigenetics, transcriptomics, and metabolomics to analyze the mechanisms of benzo[a]pyrene neurotoxicity in the hippocampus. *Toxicol. Sci., 166*, 65–81. https://doi.org/10.1093/TOXSCI/KFY192.

Wang, P., Zhang, B., Zhang, H., et al., (2019b). Metabolites change of *Scenedesmus* obliquus exerted by AgNPs. *J. Environ. Sci., 76*, 310–318. https://doi.org/10.1016/J.JES.2018.05.017.

Warth, B., Spangler, S., Fang, M., et al., (2017). *Exposome-Scale Investigations Guided by Global Metabolomics, Pathway Analysis, and Cognitive Computing* (Vol. 89, pp. 11505–11513). ACS Publ. https://doi.org/10.1021/acs.analchem.7b02759.

Wishart, D. S., (2019). NMR metabolomics: A look ahead. *J. Magn. Reson., 306*, 155–161. https://doi.org/10.1016/J.JMR.2019.07.013.

Wu, N., Yu, Y., Li, T., et al., (2016). Investigating the influence of MoS_2 Nanosheets on *E. coli* from metabolomics level. *PLoS One, 11*, e0167245. https://doi.org/10.1371/JOURNAL.PONE.0167245.

Xia, J., Bjorndahl, T. C., Tang, P., & Wishart, D. S., (2008). MetaboMiner – semi-automated identification of metabolites from 2D NMR spectra of complex biofluids. *BMC Bioinformatics, 9*. https://doi.org/10.1186/1471-2105-9-507.

Xu, Y., Wang, W., Zhou, J., et al., (2019). Metabolomics analysis of a mouse model for chronic exposure to ambient PM2.5. *Environ, Pollut., 247*, 953–963. https://doi.org/10.1016/J.ENVPOL.2019.01.118.

Yair, S., Ofer, B., Arik, E., et al., (2008). Organophosphate degrading microorganisms and enzymes as biocatalysts in environmental and personal decontamination applications. *Crit. Rev. Biotechnol., 28*, 265–275. https://doi.org/10.1080/07388550802455742.

Zhang, B., Zhang, H., Du, C., et al., (2017). Metabolic responses of the growing Daphnia similis to chronic AgNPs exposure as revealed by GC-Q-TOF/MS and LC-Q-TOF/MS. *Water Res., 114*, 135–143. https://doi.org/10.1016/J.WATRES.2017.02.046.

Zhang, H., & Zhao, L., (2017). Influence of sublethal doses of acetamiprid and halosulfuron-methyl on metabolites of zebra fish (*Brachydanio rerio*). *Aquat. Toxicol., 191*, 85–94. https://doi.org/10.1016/J.AQUATOX.2017.08.002.

Zhang, H., Huang, M., Zhang, W., et al., (2020). Silver nanoparticles alter soil microbial community compositions and metabolite profiles in unplanted and cucumber-planted soils. *Environ. Sci. Technol., 54*, 3334–3342. https://doi.org/10.1021/ACS.EST.9B07562/SUPPL_FILE/ES9B07562_SI_001.PDF.

Zhang, L. J., Qian, L., Ding, L. Y., et al., (2021). Ecological and toxicological assessments of anthropogenic contaminants based on environmental metabolomics. *Environ. Sci. Ecotechnology, 5*, 100081. https://doi.org/10.1016/J.ESE.2021.100081.

Zhang, X., Ding, X., Ji, Y., Wang, S., Chen, Y., Luo, J., ... & Peng, L. (2018). Measurement of metabolite variations and analysis of related gene expression in Chinese liquorice (*Glycyrrhiza uralensis*) plants under UV-B irradiation. *Scientific Reports*, *8*(1), 6144. https://doi.org/10.1038/s41598-018-24284-4.

Zhang, Y., Deng, Y., Zhao, Y., et al., (2014). Using combined bio-omics methods to evaluate the complicated toxic effects of mixed chemical wastewater and its treated effluent (Vol. 272, pp. 52–58). Elsevier. https://doi.org/10.1016/j.jhazmat.2014.02.041.

Zhao, L., Huang, Y., Adeleye, A. S., & Keller, A. A., (2017). Metabolomics reveals $Cu(OH)_2$ nanopesticide-activated anti-oxidative pathways and decreased beneficial antioxidants in spinach leaves. *Environ. Sci. Technol., 51*, 10184–10194. https://doi.org/10.1021/ACS.EST.7B02163.

CHAPTER 5

Plant Endomicrobiome Services in Global Food Security and Agricultural Sustainability

SHOBHIT RAJ VIMAL,[1] JAY SHANKAR SINGH,[2] and SHEO MOHAN PRASAD[1]

[1]*Ranjan Plant Physiology and Biochemistry Laboratory, Department of Botany, University of Allahabad, Prayagraj, Uttar Pradesh, India*

[2]*Department of Environmental Microbiology, School for Earth and Environmental Sciences, Babasaheb Bhimrao Ambedkar University, Lucknow, Uttar Pradesh, India*

ABSTRACT

Plant development has been positively associated with microorganisms exhibited on the surface (epiphytic), inside (endophytic) and close to the root (rhizosphere). The plant organs such as the leaf, stem, and root tissues are internally colonized by diverse and ubiquitous microorganisms comprising bacterial, fungal, archeal, and protist communities. These inside living microbes exhibited an endophytic microbiome and play a vital role in seed germination, growth, development, fitness, abiotic, pathogen protection, and diversification. The plant endomicrobiome dynamics offer mutualism to pathogenicity services for plant development. Endomicrobiome offers plant health beneficial activities such as nutrient availability, phytohormones, stress regulators, metabolites, osmoprotectants, volatile organic carbons, antibiotics, and unknown substances for growth promotion. Endophytes colonize inner host tissues, sometimes in high numbers, without a negative effect on the host and eliciting a strong defense response. The unexpected

Advances in Plant Microbiome Research for Climate Resilient Agriculture: Toward Sustainable Farming. Ashwani Kumar (Ed.)

change in a global environment, land use change, agrochemicals input, urbanization, and industrialization effectively decline crop productivity, lead to an unfertile agriculture ecosystem and affect agriculture sustainability. Poor soil health, stagnant standing crops and progressive environmental perturbations possess the problem of food insecurity. Cereal crops such as maize, rice, and wheat feed a large count of global population and ensure food and nutritional security. In this chapter, we discuss plant endomicrobiome diversity, mechanisms, services in cereal crop promotion and stress toxicity management. However, in addition, the plant endomicrobiome association offers nature-based solutions (NbS) for agricultural sustainability and next-generation global food security.

5.1 INTRODUCTION

Environmental concerns have an open door for lucrative interest in environmentally friendly sustainable agricultural technologies. The first green revolution, which relied on agrochemicals for the high yield of crops has led to environmental ruins. The present global population of about 7.2 billion is projected to pass 9.6 billion by the end of the year 2050 (United Nations, 2009). This rising population needs a vast amount to decline global hunger. The annual production of cereal crops must be revised by 50% and rises from 2.1 billion tons per year to ~3 billion tons per year. This excess crop production demand placed enormous pressure on the agriculture sector to ensure food security. The gap in crop production can be fulfilled by improvising cultivable land area and soil fertility (Vimal & Singh, 2020). The present farming sector is excessively dependent on the application of chemical synthetic fertilizers, insecticides, and pesticides, to enhance crop protection and production. These agrochemicals elevated environmental contamination, toxicity, soil health problems, infested soil, water, and air, and enhance the economy of agriculture productivity. The present world urgently required an alternative nature-based green tool to meet the present and future food demand for raising population within a very limited cultivable land area and without contaminating environmental quality. Owing to their versatility, ubiquity, and unmatchable capacity, microbial agents offer an innovative and feasible option for sustainable agriculture technology and become a possible nature-based tool for sustainable agriculture (Vimal, 2018). The microbial agent led to the developing microbiome and promises environmentally safe agriculture. Microbes efficiently colonize the plant phyllosphere, endosphere, and rhizosphere and develop associated microbiomes to influence plant growth and development through various plant growth-promoting (PGP) attributes (Vimal et al., 2017). The microbes present inside plant tissues passes whole or part of their life cycle and categorized as endophytes and offer endomicrobiome of the plant. Generally, plant-associated microbiomes facilitate plant growth directly by assisting resource acquisition or nutrient cycling or uptake, phytohormone levels and indirectly by suppression of plant pathogens or induction of resistance in host plants against pathogens (Vimal et al., 2023).

Plants began to inhabit terrestrial surfaces around 500 million years ago and were colonized by microbes for more than 407 million years on earth (Dahl & Arens, 2020). Plants are naturally associated with a variety of microorganisms in various ways and are the most studied research area of plant biology (Vimal et al., 2017). During this long term, co-evolution plant-microbe interactions has denoted by mutualism, commensalism, and parasitism phenomenon (Vimal et al., 2016). These interactions depend on environmental dynamics and microbe niche adaptability. Plant growth promoting microorganism (PGPMs) contains multiple prokaryotic and eukaryotic microbes present on the surface, inside or root zone. Root exudates offer nutrition to rhizospheric microbes and recruit beneficial microbes for plant processes (Singh et al., 2016). These microbes denoted as associate symbionts. Plant recognizes different microbes through specific receptors. These are cell surface localized pattern recognition receptors (PRRs) that identify extracellular microbe-associated molecular patterns (MAMPs) and intra-cellular nucleotide-binding leucine-rich repeat receptors (NLRs) that recognize specific microbial effector molecules. Endophytic microorganisms are hidden companions of plants living mutually beneficial life inside the host plant. Endophytes colonize a wide variety of plant tissues such as roots, stems, leaves, flowers, fruits, seeds, and pollens. Endophytes enter host tissues through either horizontal or vertical seeding methods and are colonized efficiently. Endophytes inhabit plant tissues and can contribute to host health for a long time. Endophytes have beneficial advantages such as stable pH, moisture, nutrients, water, low competition and minimum adverse environmental conditions (Singh & Vimal, 2020).

Endophyte microbiome services have been divided into three major groups: (i) nutritional quality improvement for plants; (ii) defending the host from biological stress; and (iii) abiotic stress management. The endophytes allow for to efficiently improve plant nutrition through multiple mechanisms such as nitrogen fixation, mineral solubilization, siderophore production, nutrient assimilation, phytohormones, metabolites, osmoprotectants, and other substances to ensure plant health and development. The legume crop root nodules possess several symbiotic endophytes as α-and β-rhizobia (Hardoim et al., 2015). These endophytes efficiently fix a huge amount of nitrogen and enhanced nutritional quality, soil fertility and productivity. In non-leguminous crops *Frankia*, various cyanobacteria and rhizobacteria offer nutrient recycling activity (Berg & Smalla, 2009). The eukaryotic fungal hyphae proliferate inside plant roots and regulate nutritional requirements. The mycorrhizae form vesicles and arbuscules to acquire nutrients from the soil (Vimal et al., 2017). Most endophytic microbiome services have been studied in bio-control agents.

The cereal crops are grasses that mostly belong to the family Graminae/Poaceae. The cereal crops are denoted as staple crops and grains are composed of the endosperm, germ, and bran. The continuously growing population required excess cereal grain to insecure food security and safety. The Food and Agriculture Organization of United Nations (FAO) estimates that maize (corn), wheat, and rice together account for more than three-fourths of all grain production worldwide (FAO, 2023). The FAO estimates total cereals production globally is around 2,500 million tons. Cereal supplies 51% of calories and 47% of proteins in the average diet and is recommended as nutritional food to fill food security (World Bank, 2023). The present dramatic change in global climate and other natural and anthropogenic alternations disturb crop cycle and grain productivity worldwide. The endophyte microbes can acclimatize different abiotic and biotic stress in plants and promote, protect, and enhance crop productivity to ensure food security.

5.2 THE STATE OF GLOBAL FOOD INSECURITY

The term food security was originated at World Food Conference in 1974 and defined as assuring and price stability of basic foodstuffs at the global and domestic levels. In the year 1983, FAO dictates food security as a balance between the demand and supply of food. Food security and nutrition are directly interlinked. Food insecurity can offer multiple manifestations of malnutrition (World Bank, 2023). The COVID-19 pandemic led to emphasizing the fragilities in our agro-food systems and the inequalities in our societies, driving further increases in world hunger and severe food insecurity. The food quality determines health and development of the people they eat. Food insecurity can affect diet quality in different ways, potentially leading to undernourished or defame body structure (United

Nations Food Systems Summit, 2021). Despite extensive global progress, the problem associated with poorly nourished overweight and obesity in children, maternal anemia and obesity among adults are continuously rising around the globe. In 2015, the countries of the United Nations committed to the 2030 Agenda for Sustainable Development. This agenda recognized the importance of looking beyond hunger towards the goals of ensuring access to safe, nutritious, and sufficient food for all people all year round, and of eradicating all forms of malnutrition (Sustainable Development Goals 2). The excessive enhancement in global food prices affects people's ability to afford a healthy diet (FAO, 2017). This data rises from 112 million to 3.1 billion. This count could even be even higher once data are available to account for income losses in 2020 (FAO, 2022). The Ukraine-Russia war is disrupting supply food chains and further affecting prices of grain, fertilizer, and energy. However, the world would emerge more quickly from the crisis and food security would begin to recover from such events. These man-made problems with frequent and severe extreme climatic events disrupt supply chains, especially in low-income countries.

5.3 GLOBAL CEREAL CROP PRODUCTION AND NUTRITIONAL SECURITY

During the last decade, cereal productivity overtook demand growth, leading to ample stocks and declining prices. The global trade in cereals is projected to increase by 375 Mt, to reach 3,054 Mt in 2029, mainly driven by higher yields. The balance needed in the production-stock-consumption ratio to exert downward pressure on cereal prices despite enhancement in costs (WHO, 2019). However, rapid enhancement in population requires a further vast amount of food for nutritional security. The major cereal crop includes maize in first position followed by rice and wheat presented in (Figure 5.2(a)) in the year 2022 (FAO, 2023). Maize production is projected to enhance the most (+193 Mt), followed by wheat (+86 Mt), rice (+67 Mt), and other coarse grains (+29 Mt) to ensure food and nutritional security (OECD/FAO, 2020). The global crop productivity must be enhanced by 1.1% per annum over the next decades in the limited cultivable field. This is markedly lower than the 1.9% registered in the previous decades (United Nations Food Systems Summit, 2021; FAO, 2022). The credit to agriculture data is represented in (Figure 5.1) which provides national data for over 130 countries from 2015–2021 in support of producers' cost and productivity. New biotechnology-assisted crop production and protection assured improved seed quality and microbial biotechnology-based products provide better agricultural practices. In this section, we further explore three important cereal crops rice, maize, and wheat productivity worldwide and enhancement under stress environments to ensure food security.

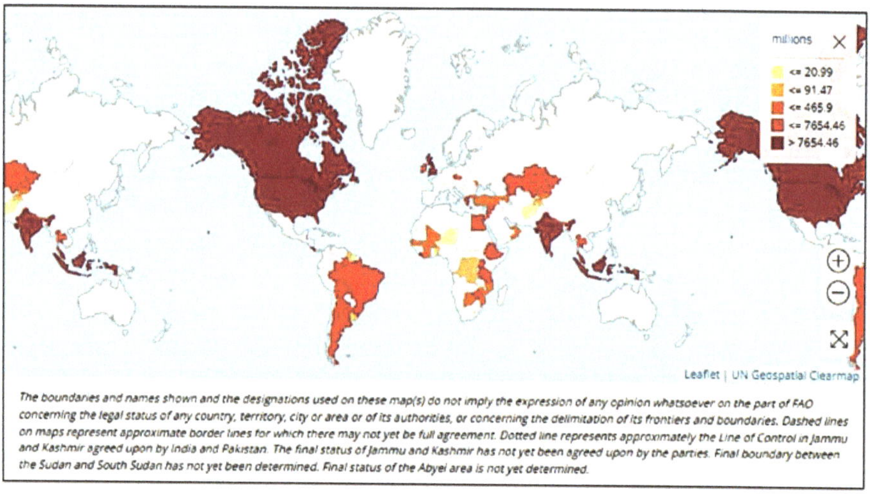

FIGURE 5.1 Credit to agriculture promotion by different parts around the globe. The credit to agriculture dataset provides national data for over 130 countries from 2015–2021 in support of producers' costs and productivity in the world agriculture sector (FAO, 2023).

5.3.1 MAIZE PRODUCTION AND ENDOMICROBIOME SERVICES

Maize (Corn) is a widely known crop in cereals for its importance as a staple food and is produced in the highest quantity among cereals (FAO, 2023). The major maize-producing countries include USA, Mexico, Nigeria, France, and Hungary. The maize belongs to the genus *Zea* and was domesticated in Southern Mexico 9,000 years ago from wild annual tropical grasses called "teosintes" with the primary ancestors being Parviglumis (*Zea mays* ssp. *parviglumis*) which are also present in present in wild (Matsuoka et al., 2002). The edible maize (*Z. mays* ssp. *mays*) domestication was bred and spread by indigenous farmers throughout America to give rise to the diverse traditional varieties. The leafy stalk of the maize plant produces pollen inflorescence and separate ovuliferous inflorescence called ears that when fertilized yield kernels or seeds, which are fruits (Xu et al., 2020). In the 21st century, scientists develop commercial inbreds and hybrids such as the temperate hybrid,

Pioneer 3751 (Smith et al., 2002). Due to this extensive potential, maize is termed a "miracle crop," and it is also entitled as the "Queen of cereals." Maize is cultivated throughout the world and a greater weight of maize is produced each year than any other grain (Moturu et al., 2022). The United States of America (USA) holds first in production amount 369 Mt, followed by China 262 Mt and Brazil 89 Mt in global data provided by FAO (Figure 5.2(b)). Wild maize has been reported to be more resistant to pests and other pathogens to grow in wild fields. The alteration in the maize genome with modern biotechnological gene editing tools improved disease susceptibility in modern maize crops.

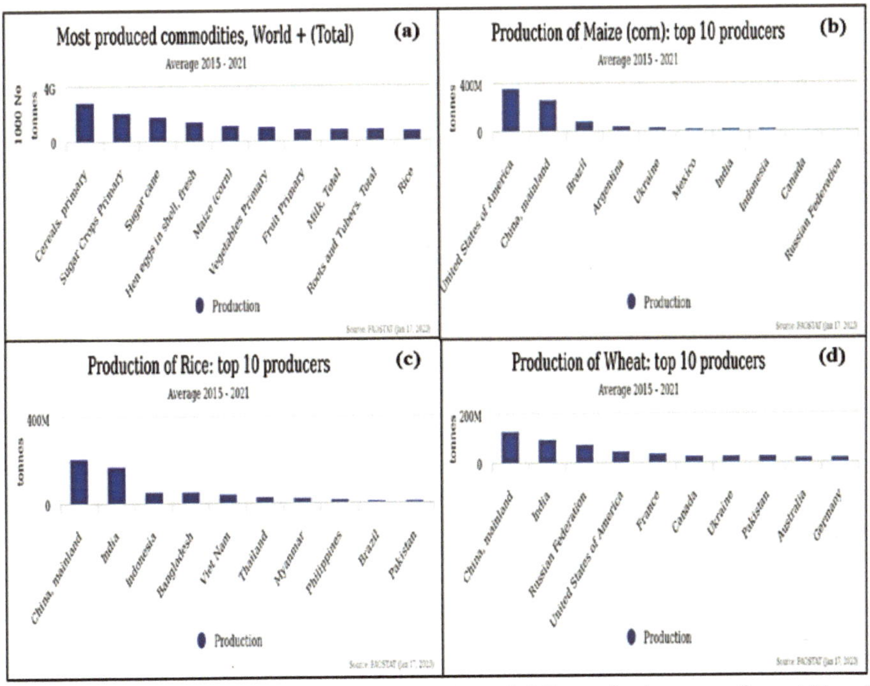

FIGURE 5.2 Worldwide quantitative production of different food sources for nutritional security between 2015 and 2021. Three important cereal crops – maize, rice, and wheat productive top 10 countries (FAO, 2023).

Maize is a vital global crop, and each seed (kernel) hosts an ecosystem of microbes living inside it (Xu et al., 2020; Moturu et al., 2022). The crop health and dynamics may change due to the association of endophytic microbes that reside inside the plants asymptomatically (Vimal & Singh, 2020). It is found that the microbes living inside plant tissue are more active and beneficial than that surface microbe (Vimal et al., 2017; Singh et al., 2016; Singh & Vimal, 2020). Endophyte microbes are always in contact with host plant tissue and directly exert a beneficial effect on plant health. The endophyte-plant association improves stress tolerance and promotes plant in different aspects. The maize crop associated with different endophytes and their beneficial activities in maize health promotion is presented in Table 5.1. Recently, Moturu et al. (2022) collected maize plants sample during the Kharif and Rabi seasons from different regions of Andhra Pradesh and isolated several potential endophytic microbes such as *Gordonia hongkongensis*, *Microbacterium hydrothermale*, *Kosakonia radicincitans*, *Kosakonia cowanii*, *Priestia megaterium*, *Priestia aryabhattai*, *Klebsiella pneumoniae*, *Cellulosimicrobium funkei*, *Bacillus licheniformis*, *Pantoea dispersa*, *Pseudomonas aeruginosa*, and *Methylorubrum populi* showing plant health promoting activities as IAA production, exopolysaccharides (EPS) and siderophore activity, solubilize mineral as (P and K), and architect potential microbiome. All these endophytes possess a potential nature-based solution for global food security and agriculture sustainability (Moturu et al., 2022). Xu et al. (2020) isolated and inoculated dark septate endophytic (DSE) fungi *Exophiala pisciphila* in maize plants and observed enhancement in plant biomass, phosphorus absorption, photosynthesis, and phosphorus enzyme activities. Inoculation changed transcriptional patterns and altered different metabolic pathways to enhance plant development (Xu et al., 2020). Ahmad et al. (2020) inoculated maize seeds with entomopathogen endophytic fungi *Metarhizium robertsii* spore and recorded enhancement in plant health, biomass, regulated defense genes and diminished pathogen insect *Agrotis ipsilon* (Ahmad et al., 2020). The maize seeds inoculated with *Yarrowia lipolytica*, isolated from spines of *Euphorbia milli* L., significantly enhanced growth parameters, controlled metabolism and enhanced endogenous IAA and declined ABA level to mitigate the negative impact of salinity stress (Gul et al., 2019). The endophytes potentially regulate phytohormones, secondary, metabolites, VOCs, and other bioactive compounds to protect and promote plant health and development. Thus, maize plant-associated endomicrobiome play vital role in crop production, nutritional security and environmental safety.

5.3.2 RICE PRODUCTION AND ENDOMICROBIOME SERVICES

Rice (Paddy) is an important staple and primary food crop grown worldwide (Vimal et al., 2019; Vimal & Singh, 2019). Rice is a vital source of nutritional security to feed almost half of the world's population, predominantly in Asia, Africa, and Latin America. Two varieties of rice *Oryza glaberrima Steud* (African rice) and *Oryza sativa* L. (Asian rice) are domesticated in different parts of the globe. The *O. glaberrima* Steud was domesticated around 2,000–3,000 years ago from the wild ancestor *Oryza barthii* by the population living in the flood plains of the upper Inland Delta of the Niger River in Mali (Linares, 2002). The morphological and molecular characterization of *O. glaberrima* assures the existence of a relatively high reservoir of genetic diversity in the area. However, without any fertilizer or biological control input African rice acquired and developed tolerance as well as resistance against multiple abiotic and biotic stresses that destroyed rice cultivation in Sub-Saharan Africa (Ndjiondjop et al., 2018). The *O. sativa* is the most common type of cultivated rice around the globe and was first domesticated in the Yangtze River basin in China. This rice variety was chiefly cultivated in temperate East Asia, upland areas of Southeast Asia, high elevated areas of South Asia, Ganga Plains and other areas throughout tropical Asia. Since 2015, China produces the most rice and is on top, in the list of FAO with 212 Mt. India is on second with 175 Mt and Indonesia with 56 Mt is the third number in the world (Figure 5.2(c)). The global rice utilization in 2021–2022 is expected to be 520.8 Mt, which is up by 1.5% from the previous season (FAO, 2021). The rice crop is highly susceptible to biotic and abiotic stress, leading to significant losses in grain yield and quality (Vimal & Singh, 2019). Different environmental stresses decline 15%–50% in global rice production. The water and saline stress majorly hamper paddy crop productivity around the globe. The stresses impair agronomic and yield-related traits of rice, including panicle development, flowering time, maturity, spikelet fertility, and grain development (Vimal et al., 2019; Vimal et al., 2020). The abiotic stress directly declines paddy crop health parameters such as root development, panicle initiation, flowering time, maturity time, spikelet fertility, grain development and weight while biotic stress leads to multiple diseases in host plants (Vimal et al., 2018; Singh & Vimal, 2020).

The paddy plants are closely associated with huge microbial diversity over plant parts and altogether represent plant microbiome. Different endophytic microbe bacteria as *Klebsiella pasteurii, Phytobacter diazotrophicus, Stenotrophomonas maltophilia, Bacillus subtilis, Pseudomonas fluorescence, Burkholderia vietnamiensis Achromobacter, Azospirillum lipoferum Pantoea*, *Flavobacterium*, *Enterobacter*, and *Curtobacterium*, *Bacillus amyloliquefaciens* and fungus *Cladosporium sphaerospermum, Paecilomyces formosus, Arthrinium phaeospermum* and other efficiently influence paddy plant health and development (Vimal & Singh, 2020; Vimal et al., 2017; Walitang et al., 2018). The paddy crops associated with different endophytes and their beneficial activities in paddy health promotion are presented in Table 5.1. These microbes significantly altered multiple plant health promotion activities as phytohormone production (Bianco et al., 2021), and stress regulators (Jaemsaeng et al., 2018) and have positive changes in plant biochemistry and physiology. The bacterial endophyte *Streptomyces* sp. positive for ACC deaminase activity significantly alleviates salinity toxicity and enhanced seed germination, chlorophyll photosynthesis capacity, K^+, Ca^+, and water contents, ROS accumulation, declined ethylene and promotes plant health (Jaemsaeng et al., 2018). The fungal endophyte isolated from the diseased plants, *T. virens* and *Aspergillus fumigatus* is effective antagonists in inhibiting the mycelial growth of *R. solani*-infected paddy crops and protecting from sheath blight disease (Motlagh et al., 2022). Thus, the different potential endophyte microbes efficiently developed endomicrobiome and facilitated plants' health and development in different ways.

5.3.3 WHEAT PRODUCTION AND ENDOMICROBIOME SERVICES

Wheat is a grass widely cultivated for its seed and is an important cereal grain and worldwide staple food. The different species of wheat together make the genus *Triticum* and *T. aestivum* L. (hexaploid bread wheat) is the most common type in the family Poaceae/Graminae (Vimal et al., 2016; Vimal et al., 2018). The global production of wheat is ranked third following rice and maize, it produces 35% of the world's total food grain. The *T. aestivum* L. is consumed by 40% of the world's population and secures nutritional security with provided 12–13.5% protein and a sizable 19% calorie (Giraldo et al., 2019). To provide nutritional and food security for a growing population, the country has to produce 140 Mt wheat by 2050 with a productivity of 4.7 Mt ha^{-1} (Sendhil et al., 2018). Wheat was first cultivated in Karacadağ mountain in southeastern turkey 9600 BC. The ancient variety tetraploid emmer wheat (*Triticum turgidum subsp. dicoccum*) was denoted as an important nutritional cereal with as high content of resistant starch, fiber, carotenoids, vitamins B2, B5, B6, A, and beneficial antioxidants. Starch is the predominant carb in the plant kingdom, accounting for over 90% of the total carb content in wheat (Suchowilska et al., 2012). Wheat can be grown in different places and environments, tropical and sub-tropical, temperate, a cold tract of the far

TABLE 5.1 Studies on the Effect of Endophytic Microbe Amendment on the World's Three Highly Productive Cereal Crops (Rice, Maize, and Wheat)

Crop	Endophytes	Host Plant	Effect of Plant Health	References
Rice	*Trichoderma virens*, *T. harzianum*, *Curvularia lunata*, *Aspergillus fumigatus*, and *Aspergillus awamori*	Diseased and healthy leaves of *Oryza sativa* L. 'Hashemi'	*T. virens* and *A. fumigatus* were found the most effective antagonists for *Rhizoctonia solani* causal agents of rice sheath blight disease.	Motlagh et al., (2022)
	Klebsiella pasteurii, *Phytobacter diazotrophicus*	African rice (*Oryza glaberrima* L.)	Inoculation significantly enhanced rice plant growth. under salinity stress.	Bianco et al., (2021)
	Stenotrophomonas maltophilia	Rice seeds	Inoculation efficiently downregulated metal stress-associated genes and promoted paddy plant health.	Zhou et al., (2019)
	Streptomyces sp.	Roots of *Clerodendrum serratum* (L.)	The strain found positive for ACC deaminase activity and efficiently declined stress ethylene levels and promoted paddy plant health.	Jaemsaeng, Jantasuriyarat, & Thamchaipenet (2018)
	Rhizobium tropici, *Rahnella* sp.	Hybrid domestic and wild poplar trees.	Endophyte-inoculated paddy plants have not exhibited (CO_2) dependent photosynthesis. The enhancement was found in photosynthetic electron transport, mesophyll conductance rate and water use efficiency.	Rho, Doty, & Kim (2020)
Maize	*Gordonia hongkongensis*, *Microbacterium hydrothermale*, *Kosakonia radicincitans*, *Kosakonia cowanii*, *Priestia megaterium*, *Priestia aryabhattai*, *Klebsiella pneumoniae*, *Cellulosimicrobium funkei*, *Bacillus licheniformis*, *Pantoea dispersa*, *Pseudomonas aeruginosa*, and *Methylorubrum populi*	Maize plant samples collected in Kharif and Rabi	These isolated efficiently produced IAA, siderophore, exopolysaccharides, solubilize P, K, and potentially promoted paddy plant health.	Moturu et al. (2022)

TABLE 5.1 *(Continued)*

Crop	Endophytes	Host Plant	Effect of Plant Health	References
	Exophiala Pisciphila, H93	Maize	The inoculated maize plants showed enhancements in biomass, phosphorus absorption, photosynthesis, and phosphorus enzyme activities. Inoculation changed the transcriptional pattern of maize related to different metabolic pathways.	Xu et al. (2020)
	Metarhizium robertsii	Plots growing winter canola (*Brassica napus* var. Wichita).	The colonized plants were improved in plant height and biomass, an altered gene involved in plant deference and insect *Agrotis ipsilon* growth suppression.	Ahmad et al., (2020)
	Yarrowia lipolytica	Spines of *Euphorbia milli* L.	The inoculation of *Y. lipolytica* had significantly promoted plant growth parameters, controlled metabolism and hormonal (ABA and IAA) and promoted maize plants under salinity stress.	Gul et al. (2019)
Wheat	*Streptomyces levis*	Cactus (*Opuntia fcus-indica*)	Inoculation enhanced seedling health and root architecture compared to control.	Govindasamy et al. (2022)
	Burkholderia ambifari	Healthy wheat plants	Inoculated wheat seeds enhanced plant growth and led to the induction of immune resistance against *Rhizoctonia cereal.*	An et al. (2022)
	Bacillus aryabhattai, B. *stratosphericus*, *Leclercia adecarboxylata*, and *Pseudomonas oryzihabitans*	Different parts of winter wheat plants.	The treated wheat seeds found variable effects on inoculated plants and significantly affect growth parameters.	Pang et al. (2022)

northern and beyond even the northern altitude. The highest amount of wheat during 2015–2021 was produced in China (mainland) 133 Mt, India 99 Mt and Russian Federation 75 Mt (Figure 5.2(d)). Wheat crops can tolerate severe cold and resume growth with the setting in of warm weather in spring.

The soil-plant-microbe symbiosis significantly affects wheat plant health and development under multiple environments. These complex interactions efficiently ameliorate the adverse impact of conventional agricultural practices and improve soil health and nutrient cycling (Govindasamy et al., 2022). Endophytes are common symbiotic bioagents such as bacteria, fungus, and actinobacteria present inside plant tissues without harming plant health. The paddy crops associated with different endophytes and their beneficial activities in paddy health promotion are presented in Table 5.1. Recently, Govindasamy et al. (2022) first time isolated endophytic actinobacteria from cactus (*Opuntia ficus-indica*). Among all 85% show PGP activities under laboratory conditions and the most efficient was identified with *Streptomyces levis*. The wheat plant isolated with this strain efficiently promotes seedling health and root architecture (Govindasamy et al., 2022). The endophyte bacteria *Burkholderia ambifari* showed antagonistic activity against *Rhizoctonia cerealis* (wheat sharp eyespot pathogenic fungus) was isolated from healthy wheat plants (An et al., 2022). The wheat seeds inoculated with *Burkholderia ambifari* significantly enhanced growth and led to the induction of immune resistance in wheat seedlings against the pathogen. The secondary metabolites (SMs) as polyphenol oxidase (PPO), peroxidase (POD), and phenylalanine-ammonia-lyase (PAL) activity were enhanced in treated wheat and involved in pathogen control mechanism. Pang et al. (2022) isolated different bacterial endophytes *Bacillus aryabhattai, B. stratosphericus, Leclercia adecarboxylata,* and *Pseudomonas oryzihabitans* from different organs (leaf, stem, and roots) of winter wheat. These isolated strains efficiently produced IAA, solubilize inorganic and organic P and grew on an N-free medium. The two ACC deaminase-producing bacteria *Kocuria rhizophila* and *Cronobacter sakazakii* inoculated wheat recorded efficient salinity stress tolerance and PGP potential (Pang et al., 2022). Thus, these entire endophytic microbes develop wheat health and are referred to as plant endomicrobiome. The beneficial associated endomicrobiome plays a vital role in plant development, productivity, and quality and ensures food safety and security.

5.4 PLANT ENDOMICROBIOME DIVERSITY AND HOST

The history of modern scientific research on beneficial microorganisms and their application in agriculture begin in 19th century. Lorenz Hiltner is recognized as the first scientist to coin the term "rhizosphere" in 1904. L. Hiltner works at the Bavarian Agriculture-Botanical Institute (later named the "Bavarian Institute of Plant Growth and Plant Protection") in Munich Germany. L. Hiltner defined the rhizosphere as a plant zone positively correlated with nutrient availability and microbial composition. Hiltner, also suggests that the quality of plant products have depend on associated flora. In the few last decades, the application of plant growth-promoting microorganisms (PGPM) attains researchers' minds due to its unique nature-based source for sustainable agriculture. The endophyte has drawn the special attention of agricultural scientists worldwide due to extensive activity to mitigate environmental stresses in crops without causing adverse effects. The "endophyte" term was denoted by De Bary in 1886, however, the fungal appearance was first recognized in 1846 by De Bary (Compant et al., 2021). The different species of three domains of life bacteria, archaea, and eukarya are reported from the endophytic association with different plants (Vimal & Singh, 2020; Vimal et al., 2023). Plant endomicrobiome is associated with diverse microflora in different organs such as roots, stems, leaves, and reproductive parts flowers and fruits, and seeds. The bacterial endophytes range from 10^5–10^7 cultivable cells per gram roots, 10^3–10^4 in steam and leaves (Compant et al., 2021). In flowers, fruits, and seeds this count ranges from 10^2–10^3 per gram of tissue (Compant et al., 2010). The pathogens successfully overcome plant defense and therefore higher in numbers than non-pathogenic hosts. The plant endomicrobiome diversity is directly correlated with host genotype, plant tissue and its vegetative stage. The endomicrobiome population is significantly affected by stress toxicity and soil characteristics (Vimal et al., 2016; Pang et al., 2022). The different plant endomicrobiome have been studied in detail in diverse geographical and climatic regions with beneficial, stress neutralizing and pathogenic activities. The endomicrobiome is composed of different bacteria, fungi, actinomycetes, and viruses reported by various authors. These endophytes have been studied for nutrient management (Xu et al., 2020), water availability (Pang et al., 2022), protection from biotic and abiotic stresses (Ahmad et al., 2020; Motlagh et al., 2022; Gul et al., 2019), growth and development (Moturu et al., 2022; Govindasamy et al., 2022), phytohormones (Jaemsaeng et al., 2018; Gul et al., 2019), SMs, VOCs, stress regulators enzymes, and exerting anti-pathogen activity (Vimal & Singh, 2020; Vimal et al., 2016; Motlagh et al., 2022; Gul et al., 2019). The wide varieties of endo microbiome-associated bio-agents significantly affected cereal crop production worldwide (Table 5.1). Multiple abiotic and biotic factors offer degrading soil characteristics, affect plant health and productivity and diminish productive ecosystem (Vimal et al., 2016; Compant et al., 2021). The various group of endophytic microbe plays a versatile role in plant health promotion and development.

The endophytic bacteria are most dominant in endomicrobiome of cereal crop plant species (Pang et al., 2022; Gul et al., 2019). The dominant different phyla include *rhizobia* in leguminous, and graminaceous plants but *Actinobacteria, Bacteroidetes*, and *Firmicutes* are actively reported as endophytes from different plants. Isrur Kruh et al. (2017) studied endophytic bacterial diversity in the parasitic weed *Phelipanche aegyptiaca* and found a clear shift from gram-negative to gram-positive bacteria at different developmental stages. They collected samples from seeds, pre-haustorium stage, tubercle, and shoot. They observed colonization of Proteobacteria (*Rhizobium, Pseudomonas, Comamonadaceae* sp., *Sphingomonas*, and *Burkholderia, Actinobacter* sp., *Bacillus* sp.) in infected and non-infected haustorium, shoot, and plant roots. The *Novosphingobium* and *Methylophilus* are found as unique endophytes for this plant species (Iasur et al., 2017). The phyla *Proteobacteria, Acidobacteria, Bacteroides, Verrucomicrobia, Planctomycetes, Chloroflexi, Firmicutes*, and *Gemmatimonadetes* are observed in grape-wine roots (Samad et al., 2017). The *Proteobacteria, Firmicutes*, and *Bacteroides* are found as dominant phyla inside maize roots (Correa-Galeote et al., 2018). Fitzpatrick & Schneider (2020) reported a unique endophytic bacterial microbiome, composition, and interaction in parasitic plant *Orobanche hederae* leaves containing phyla *Acidobacteria, Proteobacteria, Verrucomicrobia*, and bacteria belonging to *Enterobacteriaceae, Pseudomonadaceae*, and *Rhizobiaceae* (Fitzpatrick & Schneider, 2020).

Different Actinobacteria commonly referred to as "actinomycetes" Gram-positive having high G+C content in their genomes efficiently promotes plant life process. The endophytic actinomycetes are potential endophytic bio-agents that confer crop growth under abiotic and biotic stress are still inadequately understood (Vimal & Singh, 2020; Vimal et al., 2017; Compant et al., 2010). The actinomycetes comprise widespread diversity and reside inside/outside plants (*Streptomyces*), nitrogen-fixing symbionts in non-leguminous plants (*Frankia*), and plant pathogens (*Streptomyces scabies*) (Jaemsaeng et al., 2018). The endophytic actinobacterial species as Streptomyces spp., *Microbispora, Micromonospora, Nocardioides, Streptosporangium, Actinoplanes, Aeromicrobium, Arthrobacter, Brevibacterium, Corynebacterium, Microbacterium*, and some new genera *Jatrophihabitans, Herbiconiux, Jishengella, Koreibacter, Phytohabitans, Phytomonospora, Flindersiella, Actinophytocola*, and *Allonocardiopsis* were reported from different plants in plant health promotion activities (Govindasamy et al., 2022). These endophyte actinomycetes exhibited multiple physiological and biochemical characteristics and produce extracellular enzymes, antibiotics, and a wide variety of SMs. In earlier years endophytic actinomycetes gain a wide variety in abiotic and biological control activities.

The fungal endophytes play a key role in the development and habitat adaptation of plants under multiple biotic and abiotic stresses (Vimal et al., 2017; Vimal et al., 2023). The plants are natural symbionts of fungal endophytes and/or mycorrhizal fungi since the discovery of Deberry (Fitzpatrick & Schneider, 2020). The fungal endophytes produced novel SMs, antibiotics, vitamins, and volatile organic carbons (Gul et al., 2019). Different fungal endophytes play the versatile role of antagonistic activity and protect cereal crops from different pathogens and also improved the management of post-harvest control (Govindasamy et al., 2022). The two groups of endophytic fungi clavicipitaceous endophytes (C-endophytes) found in grasses and the non-clavicipitaceous endophytes (NC-endophytes) colonized asymptomatic tissues of non-vascular plants, ferns and allies, conifers, and angiosperms are studied. The C-endophytes denoted a small number of phylogenetically related clavicipitaceous species, build symbiotic associations inside plant shoots and are limited to some cool and warm season grasses (De Bary, 1879). The NC-endophytes are exceedingly diverse and represent a polyphyletic assemblage of primarily ascomycetous fungi with different ecological roles. The NC-endophytes have differentiated into three functional classes based on host colonization, transmission between host generations, planta diversity levels, and different ecological function (Rodriguez et al., 2009). The Clavicipitaceae (Hypocreales; Ascomycota) include free-living and symbiotic species associated with or grasses, rushes, and sedges. The family is relatively derived within the Hypocreales in which many produced multiple bioactive compounds. The NC-endophytes represent members of the Dikarya (*Ascomycota* or *Basidiomycota*). Most belong to the Ascomycota, with a minority of *Basidiomycota* (Arnold, 2007). The *Basidiomycota* include some members of the *Agaricomycotina* and *Pucciniomycotina*. The arbuscular mycorrhizal fungi (AMF), *Glomeromycotina*, and *Mucoromycotina*, colonized the plant's roots as fine endophytes and provide multiple support at different plant developmental stages (Vimal & Singh, 2020). The functional diversity in these symbioses depends on both fungal and plant life histories and is influenced by the environment. Thus, these microbial endophytes' symbiosis develops plant endomicrobiome and provide multiple services in plant development.

5.5 ENDOMICROBIOME SERVICES STRESS MANAGEMENT AND AGRICULTURE SUSTAINABILITY

The endophytes efficiently colonize plant tissues and develop symbiotic or mutualistic interactions with host plants. They interact and communicate with the host plant more efficiently than rhizospheric colonizers (Vimal et al., 2023; Vimal et al., 2020). In addition, endophytes get stable equilibrium for the function to compare to variation in soil conditions. The root nodules host specific microbe and facilitate mineral and enhance soil fertility equilibrium. The endophyte attracts a good deal of research interest due to its ability to nurture host plants with different services (Harman & Uphoff, 2019). These services include phytohormone regulation, stress modulators, mineral nutrients (N, P, and K) availability, SMs, antibiotics, and VOCs (Moturu et al., 2022; Rodriguez et al., 2009).

Different endophytes associated endomicrobiome services in abiotic, biotic stress and nutrient and fertility management are represented in Figure 5.3. The diazotrophs residing in root nodules are efficient in atmospheric nitrogen fixation in leguminous and non-leguminous plants. Different endophytic diazotrophs efficiently fix atmospheric nitrogen in different crops. Phosphorus has been solubilized by different endophytes (Vimal et al., 2016). The endophytes secret organic acids that are considered to convert inorganic phosphate into a soluble form and are available to plants (Moturu et al., 2022). The phytohormone IAA is the most prevalent entrant of the Auxin family and significantly controlled overall plant physiology affecting cell division, differentiation, and tropism (gravity and light). Different endophytes can produce IAA and promote plant growth under nutrient-limited conditions (Singh et al., 2016; Hardoim et al., 2015). IAA acts as regulating phytohormone in seedling germination, cotyledon development and different growth stage in plant health. IAA efficiently affects cell development, spore germination and mycelial elongation. The biosynthetic pathway of IAA is similar in plants and bacteria and tryptophan plays a precursor role (Berg & Smalla, 2009; Harman & Uphoff, 2019). Different endophytic isolates have been reported around cereal crop production enhancement around the globe (Bianco et al., 2021; Motlagh et al., 2022). Ethylene plays a vital role in plant development under normal conditions but under stress, peaks rise in the graph. Ethylene is important during the initial phase of plant development and is essentially required by some plants for seed germination and it was observed that the rate of ethylene production increases during germination and seedling development (Abeles et al., 1992). The abnormal hike in ethylene peak was noted in response to plants under abiotic as salinity, metals, chemicals, temperature, water, ultra-violet light, and biotic as insects, nematodes, and phytopathogens (bacteria and fungi) damaged plants. The higher ethylene level causes senescence, chlorosis, and leaf abscission (Glick, 2014). The IAA-induced transcription of plant enzyme ACC synthase catalyzes the formation of 1-aminocyclopropane-1-carboxylic acid (ACC). This ACC metabolizes and enhanced plant ethylene levels under stress environments (Vimal et al., 2017). The enzyme ACC deaminase is also known as a stress regulator and the decline in the level of stress-induced ethylene was first characterized in 1978 by Honma & Shimoura (1978). This enzyme was isolated from the bacteria *Pseudomonas* sp. strain ACP and fungus *Penicillium citrinum* and some microbes have been reported for stress amelioration in different crops around the globe (Honma & Shimomura, 1978; Honma, 1993). These ACC deaminase-positive microbes cleaved the plant ethylene precursor, ACC, to produce ammonia and α-ketobutyrate and decline the negative effect of stress toxicity (Vimal et al., 2019). The rhizobium strains exhibit ACC deaminase activity and are >40% highly efficient at forming n-fixing nodules that lack activity (Ma et al., 2004).

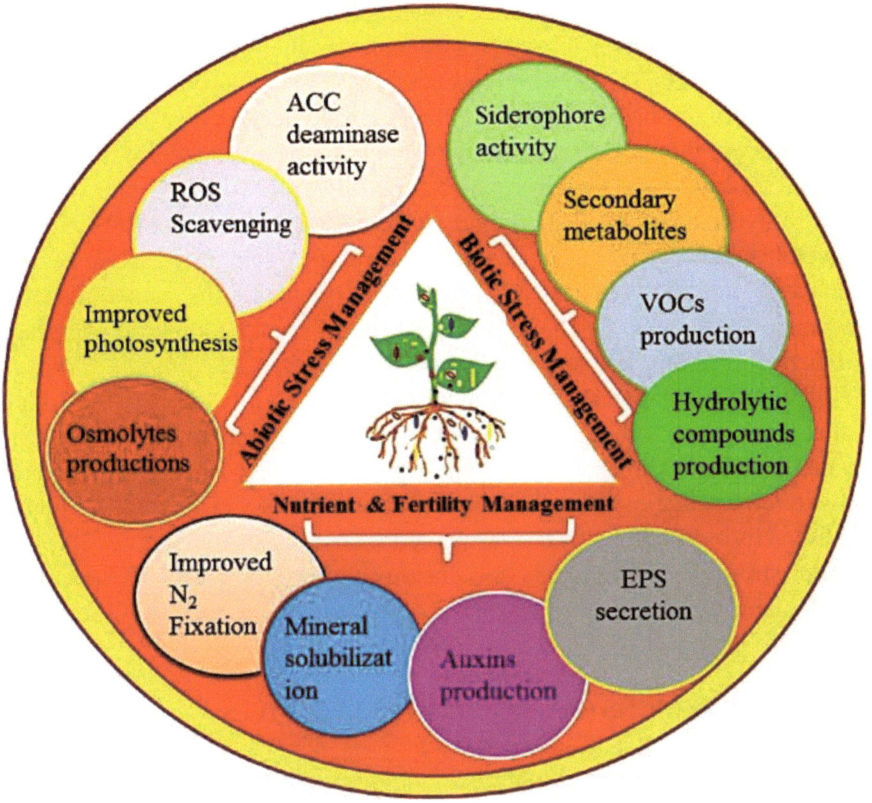

FIGURE 5.3 Different endophytes associated endomicrobiome services in abiotic, biotic stresses, nutrient, and soil fertility management to enhance crop productivity and ensure food security.

In biotic stress management or biological control, endophyte-associated endomicrobiome plays a versatile role in the antagonistic activity and pathogenesis of multiple pathogens. The plant rhizomicrobiome first suppresses the soil-borne pathogens and is denoted as the first microbiological layer of plant defense (Vimal et al., 2017; Vimal et al., 2023). However, some pathogenic microbes bypass this first barrier and establish physical interaction with host cortical cells. These infections triggered the innate mechanism of plant defense and inner endomicrobiome against pathogens (Compant et al., 2021). The pant endomicrobiome act as the second layer of defense and protect the plant from multiple pathogens (Rodriguez et al., 2009). Different phyto-pathogens *Collectotrichum* sp., *Phytoththora* sp., *Fusarium* sp., *Rhizoctonia* sp., have been described to be inhabited by different endophytic microorganisms (An et al., 2022; Gul et al., 2019). These microbes produce different compounds as biomacromolecules as thermostable enzyme lipase (Torres et al., 2003), and amylopectin hydrolyzing *glucoamylase* produced by *Acremonium* sp. (Marlida et al., 2000). Different SMs as alkaloids (amines and amides, indole derivatives, pyrrolizidines, and quinazolines), steroids (20S,22S-4a-homo-22-hydroxy-4-oxaergasta-7,24(28)-dien-3-one and 1873b-hydroxyergosta-4,22-diene), terpenes (sesquiterpenoids), Quinones (cyclohexenone epoxides, jesterone, ambuic acid, hydroxyjesterone), Phenols, and polyphenols (1,3-dihydro isobenzofurans, orsellinic acid, esersglobosumones) has been produced by different endophytic bacteria as *Aspergillus fumigatus,* and fungi *Gliocaldium* sp., *Chaetomium globosum, Pestalotiopsis* sp. and others (Ahmad et al., 2020; Gul et al., 2019; Motlagh et al., 2022). Low molecular weight iron-chelating molecules denoted as a siderophore. The microbes efficiently produced siderophores have antagonistic activity against fungal pathogens (Govindasamy et al., 2022). Thus, all these activities are positively associated with plant health and development under stressful environments and ensure food productivity and agricultural sustainability.

5.6 CONCLUSIONS AND FUTURE PERSPECTIVES

The current recessionary situation around the globe due to the early microbe-mediated pandemic led to challenges for governments to enhance the agricultural budget for agro-food system transformation. In the future, cereal crops are a vital source of nutritional and food security for the escalating population. The ongoing increase in population, and multiple natural and anthropogenic sources have affected cereal crop production worldwide. The above cereal crop production data of FAO-Switzerland exhibit the additional requirement of cereal production for the next generation population in upcoming decades. The negative effects of agrochemicals offer plant endomicrobiome services as potential bio-tool to enhance crop health, and productivity and ensure nutritional security and agricultural sustainability. Different active microbes such as bacteria, fungi, and actinomycetes in endomicrobiome possess multiple plant-growth-promoting and antagonistic activities and enhance cereal crop production. The manipulation of plant endomicrobiome with efficient strains led to the development of nature-based tools for sustainable crop production. The engineered seed microflora offers plant health development and biological control and are a vital source for future crop output. The endomicrobiome services offer a cost-effective and efficient natural tool for low-cost nutritious food production. Thus, a detailed study is required to explore plant-beneficial endomicrobiome and its application in cereal crop production and protection to feed the next generation population.

ACKNOWLEDGMENT

SRV is thankful to University Grants Commission, New Delhi India for financial support in form of Dr. D. S. Kothari Postdoctoral Fellowship (Award letter No. F.4-2/2006 (BSR)/BL/20-21/393 (September 14, 2021) 90th List).

DECLARATION OF COMPETING INTEREST

The authors declare that they have no conflict of interest.

KEYWORDS

- **agriculture sustainability**
- **endomicrobiome**
- **endophytes**
- **food security**
- **global food insecurity**
- **microorganisms**
- **nutritional security**

REFERENCES

Abeles, F. B., Morgan, P. W., & Saltveit, M. E. Jr., (1992). *Ethylene in Plant Biology* (2nd edn.). New York: Academic Press.

Ahmad, I., Jiménez-Gasco, M. M., Luthe, D. S., Shakeel, S. N., & Barbercheck, M. E., (2020). Endophytic *Metarhizium robertsii* promotes maize growth, suppresses insect growth, and alters plant defense gene expression. *Biological Control, 144*, 104167.

An, C., Ma, S., Liu, C., Ding, H., & Xue, W., (2022). *Burkholderia ambifaria* XN08: A plant growth-promoting endophytic bacterium with biocontrol potential against sharp eyespot in wheat. *Frontiers in Microbiology, 13*, 1–11.

Arnold, A. E., (2007). Understanding the diversity of foliar fungal endophytes: Progress, challenges, and frontiers. *Fungal Biology Reviews, 21*, 51–66.

Bary, A., (1866). *Morphologie und Physiologie der Pilze, Flechten und Myxomyceten.* W Engelmann.

Berg, G., & Smalla, K., (2009). Plant species and soil type cooperatively shape the structure and function of microbial communities in the rhizosphere. *FEMS Microbiology Ecology, 68*, 1–13.

Bianco, C., Andreozzi, A., Romano, S., Fagorzi, C., Cangioli, L., Prieto, P., Cisse, F., et al., (2021). Endophytes from African rice (*Oryza glaberrima* L.) efficiently colonize Asian rice (*Oryza sativa* L.) stimulating the activity of its antioxidant enzymes and increasing the content of nitrogen, carbon, and chlorophyll. *Microorganisms, 9*, 1–27.

Bischoff, J. F., & White, J. F. Jr., (2005). Evolutionary development of the *Clavicipitaceae*. In: Dighton, J., White, J. F., & Oudemans, P., (eds.), *The Fungal Community: Its Organization and Role in the Ecosystem* (pp. 505–518). Boca Raton, FL, USA: Taylor & Francis.

Compant, S., Carbon, M. C., Vacher, C., Mitter, B., Samad, A., & Sessitsch, A., (2021). The plant endosphere world – bacterial life within plants. *Environmental Microbiology, 23*, 1812–1829.

Compant, S., Clement, C., & Sessitsch, A., (2010). Plant growth-promoting bacteria in the rhizo- and endosphere of plants: Their role, colonization, mechanisms involved and prospects for utilization. *Soil Biology Biochemistry, 42*, 669–678.

Correa-Galeote, D., Bedmar, E. J., & Arone, G. J., (2018). Maize endophytic bacterial diversity as affected by soil cultivation history. *Frontiers Microbiology, 9*, 1–9.

Dahl, T. W., & Arens, S. K. M., (2020). The impacts of land plant evolution on Earth's climate and oxygenation state – An interdisciplinary review. *Chemical Geology, 547*, 1–25.

De Bary, A., (1879). Die Erscheinung der symbiose. In: Trubner, K. J., (ed.), *Vortrag auf der Versammlung der Naturforscher und Ärtze zu Cassel* (pp. 1–30). Strassburg, Germany: Verlag.

FAO, (2017). *Strengthening Sector Policies for Better Food Security and Nutrition Results: Political Economy Analysis*. Policy Guidance Note 8. Rome. www.fao.org/3/i7212en/I7212EN.pdf (accessed on 21 October 2023).

FAO, (2021). *COVID-19: Agricultural Trade and Policy Responses During the First Wave of the Pandemic in 2020*. FAO support to the WTO negotiations at the 12th Ministerial Conference. Trade Policy Brief No. 37. Rome, FAO. https://doi.org/10.4060/cb5406en.

FAO, (2022). *World Food and Agriculture – Statistical Yearbook 2022*. Rome.

FAO, (2023). *FAOSTAT: Credit to Agriculture*. In: FAO. Rome. Map geodata [shapefiles]. New York, USA, UN. https://www.fao.org/faostat/en/#data/IC/visualize (accessed on 21 October 2023).

Fitzpatrick, C. R., & Schneider, A. C., (2020). Unique bacterial assembly, composition, and interactions in a parasitic plant and its host. *Journal Experimental Botany, 71*, 2198–2209.

Giraldo, P., Benavente, E., Manzano-Agugliaro, F., & Gimenez, E., (2019). Worldwide research trends on wheat and barley: A bibliometric comparative analysis. *Agronomy, 9*, 1–18.

Glick, B. R., (2014). Bacteria with ACC deaminase can promote plant growth and help to feed the world. *Microbiology Research, 169*, 30–39.

Govindasamy, V., George, P., Ramesh, S. V., Sukreshkumar, P., Rane, J., & Minhas, P. S., (2022). Characterization of root-endophytic actinobacteria from cactus (*Opuntia fcus-indica*) for plant growth promoting traits. *Archives of Microbiology, 204*, 1–14.

Gul, J. F., Hamayun, M., Hussain, A., Jan, G., Iqbal, A., Khan, A., & Lee, I. J., (2019). An endophytic isolate of the fungus *Yarrowia lipolytica* produces metabolites that ameliorate the negative impact of salt stress on the physiology of maize. *BMC Microbiology, 19*, 1–10.

Hardoim, P. R., van-Overbeek, L. S., Berg, G., Pirttila, A. M., Compant, S., Campisano, A., Doring, M., & Sessitsch, A., (2015). The hidden world within plants: Ecological and evolutionary considerations for defining functioning of microbial endophytes. *Microbiology and Molecular Biology Reviews, 79*, 293–320.

Harman, G. E., & Uphoff, N., (2019). Symbiotic root-endophytic soil microbes improve crop productivity and provide environmental benefits. *Scientifica*, 1–25.

Honma, M., & Shimomura, T., (1978). Metabolism of 1-aminocyclopropane-1-carboxylic acid. *Agricultural Biological Chemistry, 43*, 1825–1831.

Honma, M., (1993). Stereospecific reaction of 1-aminocyclopropane-1-carboxylate deaminase. In: Pech, J. C., Latche, A., & Balague, C., (eds.), *Cellular and Molecular Aspects of the Plant Hormone Ethylene* (pp. 111–116). Dordrecht, The Netherlands: Kluwer Academic Publishers.

Iasur, K. L., Lahav, T., Abu-Nassar, J., Achdari, G., Salami, R., Freilich, S., & Aly, R., (2017). Host-parasite-bacteria triangle: The microbiome of the parasitic weed phelipanche aegyptiaca and tomato-*Solanum lycopersicum* (Mill.) as a host. *Frontiers Plant Science, 8*, 1–9.

Jaemsaeng, R., Jantasuriyarat, C., & Thamchaipenet, A., (2018). Molecular interaction of 1-aminocyclopropane-1-carboxylate deaminase (ACCD)-producing endophytic *Streptomyces* sp. GMKU 336 towards salt-stress resistance of *Oryza sativa* L. cv. KDML105. *Scientific Reports, 8*, 1–5.

Linares, O. F., (2002). African rice (*Oryza glaberrima*): History and future potential. *Proc. Natl. Acad. Sci. USA, 99*, 16360–16365.

Ma, W., Charles, T. C., & Glick, B. R., (2004). Expression of an exogenous 1-aminocyclopropane- 1-carboxylate deaminase gene in *Sinorhizobium meliloti* increases its ability to nodulate alfalfa. *Applied Environmental Microbiology, 70*, 5891–5897.

Marlida, Y., Saari, N., Hassan, Z., & Radu, S., (2000). Raw starch degrading enzyme from isolated strains of endophytic fungi. *World Journal Microbiology Biotechnology, 16*, 573–578.

Matsuoka, Y., Vigouroux, Y., Goodman, M. M., Sanchez, J., Buckler, E., & Doebley, J., (2002). A single domestication for maize shown by multilocus microsatellite genotyping. *Proceedings of the National Academy of Sciences United States of America., 99*, 6080–6084.

Motlagh, M. R. S., Jahangiri, B., Kulus, D., Tymoszuk, A., & Kaviani, B., (2022). Endophytic fungi as potential biocontrol agents against *Rhizoctonia solani* J.G. Kühn, the causal agent of rice sheath blight disease. *Biology, 11*, 1–15.

Moturu, U. S., Nunna, V., Avula, V. G., Jagarlamudi, V. R., Gutha, R. R., & Tamminana, S., (2022). Investigating the diversity of bacterial endophytes in maize and their plant growth-promoting attributes. *Folia Microbiologica.* https://doi.org/10.1007/s12223-022-01015-x.

Ndjiondjop, M. N., Wambugu, P., Sangare, J. R., Dro, T., Kpeki, B., & Gnikoua, K., (2018). *Oryza glaberrima Steud.* In: Mondal, T., Henry, R., (eds.), The *Wild Oryza* Genomes; *Compendium of Plant Genomes* (105–126). Springer: Cham, Switzerland.

OECD/FAO, (2020). *OECD-FAO Agricultural Outlook*. OECD Agriculture statistics (database). http://dx.doi.org/10.1787/agr-outl-data-en.
Pang, F., Tao, A., Ayra-Pardo, C., Wang, T., Yu, Z., & Huang, S., (2022). Plant organ- and growth stage-diversity of endophytic bacteria with potential as biofertilizers isolated from wheat (*Triticum aestivum* L.). *BMC Plant Biology, 22*, 1–16.
Rho, H., Doty, S. L., & Kim, S. H., (2020). Endophytes alleviate the elevated CO_2-dependent decrease in photosynthesis in rice, particularly under nitrogen limitation. *J. Experimental Botany, 71*, 707–718.
Rodriguez, R. J., White, Jr. J. F., Arnold, A. E., & Redman, R. S., (2009). Fungal endophytes: Diversity and functional roles. *New Phytologist, 182*, 314–330.
Samad, A., Trognitz, F., Compant, S., Antonielli, L., & Sessitsch, A., (2017). Shared and host-specific microbiome diversity and functioning of grapevine and accompanying weed plants. *Environmental Microbiology, 19*, 1407–1424.
Sendhil, R., Kumar, T. M. K., & Singh, G. P., (2019). Wheat production in India: Trends and prospects. In: Shah, F., Khan, Z., Iqbal, A., Turan, M., & Olgun, M., (eds.), *Recent Advances in Grain Crops Research* (pp. 1–16). London: Intech Open Limited.
Singh, J. S., & Vimal, S. R., (2020). *Microbial Services in Restoration Ecology* (pp. 157–179). Elsevier Netherlands.
Singh, J. S., Kaushal, S., Kumar, A., Vimal, S. R., & Gupta, V. K., (2016). Book review: Microbial inoculants in sustainable agricultural productivity-vol. II: Functional application. *Frontiers Microbiology, 7*, 1, 2.
Smith, J. S. C., Duvick, D. N., Smith, O. S., Cooper, M., & Feng, L., (2002). Changes in pedigree backgrounds of Pioneer brand maize hybrids widely grown from 1930 to1999. *Crop Science, 44*, 1935–1946.
Suchowilska, E., Wiwart, M., Kandler, W., & Krska, R., (2012). A comparison of macro- and microelement concentrations in the whole grain of four *Triticum* species. *Plant Soil Environment, 58*, 141–147.
Torres, M., Dolcet, M. M., Sala, N., & Canela, R., (2003). Endophytic fungi associated with Mediterranean plants as a source of mycelium-bound lipases. *Journal Agricultural Food Chemistry, 51*, 3328–3333.
UN (United Nations), (2009). *The Sustainable Development Goals Report, 64*.
United Nations Food Systems Summit, (2021). *The Coalition to Repurpose Public Support to Food and Agriculture*. In: United Nations Food Systems Summit 2021. https://www.un.org/en/food-systems-summit (accessed on 21 November 2023).
Vimal, S. R., & Singh, J. S., (2019). Salt tolerant PGPR and FYM application in saline soil paddy agriculture sustainability. *Climate Change Environmental Sustainability, 7*, 61–71.
Vimal, S. R., & Singh, J. S., (2020). Microbial services to nurture plant health under stressed soils. In: Singh, J. S., & Vimal, S. R., (eds.), *Microbial Services in Restoration Ecology* (pp. 157–179). Elsevier Netherlands.
Vimal, S. R., (2018). *Effect of Plant Growth Promoting Rhizobacteria (PGPR) and Farmyard Manure (FYM) Amendment on Growth Parameters and Antioxidant Level in Paddy (Oryza sativa L.) Crop Under Soil Salinity*. Ph.D. thesis. Babasaheb Bhimrao Ambedkar University, Lucknow, Uttar Pradesh, India. http://14.139.228.232:8080/jspui/handle/0522/419.
Vimal, S. R., Gupta, J., & Singh, J. S., (2018). Effect of salt tolerant *Bacillus* sp. and *Pseudomonas* sp. on wheat (*Triticum aestivum* L.) growth promotion under soil salinity: A comparative study. *Microbiology Research, 9*, 26–32.

Vimal, S. R., Patel, V. K., & Singh, J. S., (2019). Plant growth promoting *Curtobacterium albidum* strain SRV4: An agriculturally important microbe to alleviate salinity stress in paddy plants. *Ecological Indicators, 105*, 553–562.

Vimal, S. R., Singh, J. S., & Prasad, S. M., (2023). Plant-microbe dynamics: Nature based solutions for sustainable agriculture. *Anthropocene Science*. (Accepted MS No. ANPS-D-22-00064). https://doi.org/10.1007/s44177-023-00043-7.

Vimal, S. R., Singh, J. S., Arora, N. K., & Singh, D. P., (2016). PGPR-an effective bio-agent in stress agriculture management. In: Sarma, B. K., & Jain, A., (eds.), *Microbial Empowerment in Agriculture-A Key to Sustainability and Crop Productivity* (pp. 81–108). Biotech Books Publishing Agency, New Delhi.

Vimal, S. R., Singh, J. S., Arora, N. K., & Singh, S., (2017). Soil plant microbe interactions in stressed agriculture management: A review. *Pedosphere, 27*, 177–192.

Vimal, S. R., Yadav, S., Rai, P. K., & Singh, J. S., (2020). Role of *Bacillus humi* (SS4) in management of legume *Vigna radiata* (L.) stresses under saline environments. *Climate Change Environmental Sustainability, 8*, 54–66.

Walitang, D. I., Kim, C. G., & Kim, K., (2018). The influence of host genotype and salt stress on the seed endophytic community of salt-sensitive and salt-tolerant rice cultivars. *BMC Plant Biology, 18*, 1–16.

WHO, (2019). *Essential Nutrition Actions: Mainstreaming Nutrition Through the Life-Course*. Geneva, Switzerland.

World Bank, (2023). https://www.worldbank.org/en/programs/icp/brief/foodpricesfornutrition (accessed on 21 October 2023).

Xu, R., Li, T., Shen, M., Yang, Z. L., & Zhao, Z. W., (2020). Evidence for a dark septate endophyte (*Exophiala Pisciphila*, H93) enhancing phosphorus absorption by maize seedlings. *Plant & Soil., 452*, 249–266.

Zhou, J., Li, P., Meng, D., Gu, Y., Zheng, Z., & Yin, H., (2019). Isolation, characterization and inoculation of Cd tolerant rice endophytes and their impacts on rice under Cd contaminated environment. *Environmental Pollution, 260*, 1–9.

CHAPTER 6

Future Perspectives of Endophytic Microbiome in Sustainable Agriculture

KANIKA CHOWDHARY, HIMANSHU ARORA, and SATYAWATI SHARMA

Indian Institute of Technology, CRDT, Hauz Khas, Delhi, India

ABSTRACT

Endophytes are biotrophic microbes, colonizing the host plant in a mutually symbiotic relationship. Endophytes gain entry and colonize the plant cells by overturning the triggered plant defense mechanism. Endophytes synthesize a plethora of secondary metabolites (SMs), many of which are structurally and chemically similar to the host plant, which has been attributed to the *de facto* genetic interchange between the endophyte and the host. The advancement in biotechnological approaches has opened up new domains and has provided insights into the life cycle of endophytes and their potential use in sustainable agriculture. Formulations based on biological control agents have huge market potential. A top-down approach in the isolation and screening of biological agent provides a more suitable and efficient candidate microbe, which have survived multiple stresses in actual field conditions. Agricultural practices have a direct and very significant effect on microbiomes and endophytic diversity. Sustainable agriculture can be promoted using organic alternatives of fertilizers and pesticides which encourage the microbiome health and diversity. *Piriformospora indica* is an endophyte with a wide host range and is known to have photoprotective and phytopromotional capabilities, holding immense potential in encouraging sustainable agriculture.

Advances in Plant Microbiome Research for Climate Resilient Agriculture: Toward Sustainable Farming.
Ashwani Kumar (Ed.)

6.1 INTRODUCTION: PLANT HOLOBIONT

Plants are multitrophic, dynamic, adaptable, and ever-evolving habitat systems continuously acclimatizing themselves in response to various surrounding changes. Plant's associated microbial assemblage is known as "microbiome." Renowned authors have described microbiome as genetic pool inhabited within a macro-organism in a given environment at a particular time. The microbiome collaborates in a dynamic fashion to create synergistic relationships by influencing the host's physiology (Chowdhary & Kaushik, 2019a).

When an endophyte gains entry into the host plant tissue, it allegedly initiates a cascade of signal molecules in the host plant. All invading microbes regardless of their pathogenic levels induce MAMP (microbe/pathogen-associated molecular patterns) – elicited plant defense mechanism known as pattern-triggered immunity (PTI). Endophytic microbes pre-emptively protect themselves by overturning the immune system of the plant, such as by producing chitin deacetylases which degrades cell wall chitin into oligomers that evade plants' receptor (R) proteins endophyte-mediated resistance (EMR). In a recent study, *Fusarium oxysporum* induced EMR in tomato plants in which PTI signaling is compromised (de Lamo et al., 2021).

Endophytic microbes have a higher ability to colonize host plants than contemporary pathogens, often inhibiting pathogen's growth and modulating the nutrient balance in a way that favors endophyte's existence. Plants have a multidimensional range of endophytic-pathogenic lifestyles. They are rather transposable in nature and are generally impacted by outward physiological, chemical, and molecular factors. Based on the lifestyle selected by endophytes inside the plant host, they are largely categorized into obligate, facultative, and passive lifestyles. Obligate endophytes do not reproduce outside the host tissue and mostly get transferred horizontally via seeds and seedlings. Whereas facultative endophytes get transmitted from the ground and can persist independently from the plant host, most of the time shuttle between their host plant and surrounding environment. Most of the plant-growth-promoting fungal endophytes are facultative in nature. Passive endophytes are those microbes that do not purposely inhabit host plants and gain access into plant tissues through injuries and others. For instance, root endophytic fungus *Piriformspora indica* was found to contain genetic make-up for both endophytic and saprophytic lifestyles as described by comparative multigenomic investigation (Gill et al., 2016).

6.2 BIOACTIVE METABOLITES STUDIED/EXAMINED IN FUNGAL ENDOPHYTES

Filamentous fungi have been recognized to produce > 45,000 secondary metabolites (SMs) having enormous biological activities (El-Sayed et al., 2017). Various studies have described that the synthesis of SMs involves only a few clusters of genes, which are duly expressed during conventional fermentation techniques. Endogenous hypothesis supports that because endophytes originate from the host plant's mitochondria and chloroplast and henceforth their genetic build-up is similar to that of the host. The entire spectrum of SMs synthesized in micro-fungi utilizes a few precursor compounds such as polyketides, alkaloids, terpenes, and non-ribosomal peptides. Non-ribosomal peptides are synthesized by enzymatic complex termed as nonribosomal peptide synthetases (NRPS). The competency of endophytes to produce chemically and structurally similar metabolites as that of the host plant has been contemplated to occur due to the *de facto* genetic interchange between endophytes and their respective host. Being able to exist in a highly functional niche, fungal endophytes tend to remain in a metabolically aggressive state thereby synthesizing an inimitable array of metabolites (Chowdhary & Sharma, 2020) (Table 6.1).

6.3 ROLE OF MICROBES IN SUSTAINABLE AGRICULTURE

Agriculture and its allied sector are a significant foundation stone of the growing Indian economy and it accounts for about 20% of the country's GDP. Approximately 70% of the rural families are dependent on agriculture as the primary source of livelihood. Indian agriculture is hugely dependent on nature's gamble since 2/3 of agricultural land is in rained water and utilizes rainwater as the main source of irrigation. It is, however, indubitable that chemical pesticides have caused a considerable increase in productivity and added significantly to the agricultural industry, nevertheless, pesticides pose severe apprehensions. Such chemicals are responsible for health threats to end-users during working and application of pesticides in treated lands. Inaccurate application, inappropriate storage, reuse of contaminated containers for consuming edible materials could cause accidental and unintended poisonings. Often, pesticide applications incidentally cause damage to beneficiary natural enemies of pathogens and pests and increase the growing likelihood of pesticide resilient species. Additionally, well-informed farmers

TABLE 6.1 Management of Different Pathogen and Pest by Fungal Endophytes

Sl. No.	Endophytic Fungi	Bioactivities Extended to Host Plant Against Plant Pathogens/Pest	References
1.	*Phialocephala scopiformis*	Insecticidal bioactive molecule (rugulosin) towards *Choristoneura fumiferana*.	Larkin et al. (2012)
2.	*Pestalotiopsis virgulata*	Mosquitocidal action towards *Aedes aegypti*.	Bücker et al. (2013)
3.	*Mycoleptodiscus indicus*	Mosquitocidal action towards *Aedes aegypti*.	Rosa et al. (2012)
4.	*Beauveria* KL042, *Paecilomyces* KL017 and *Alternaria* KL062	Insecticidal activity.	Hong et al. (2010)
5.	*Macrophomina phaseolina*	Antifungal activity against *Sclerotinia sclerotiorum*.	Chowdhary & Kaushik (2019b)
6.	*Daldinia cf. concentrica*	Antifungal (*A. niger*) activity in peanuts by VOCs.	Liarzi et al. (2016)
7.	*Nodulisporium* sp. *GS4d2II1a*	Antifungal action towards *Pythium aphanidermatum* by VOCs.	Sánchez-Fernández et al. (2016)
8.	*Geotrichum candidum*	Antifungal action towards *Rhizoctonia solani* by VOCs.	Mookherjee et al. (2018)
9.	*Daldinia eschscholtzii*	Antifungal action towards *Colletotrichum acutatum* by VOCs.	Khruengsai et al. (2021)
10.	*Simplicillium aogashimaense*	Antifungal action against *Diplodia corticola* and *Biscogniauxia mediterranea* by VOCs.	Costa et al. (2020)
11.	*Trichoderma atroviride*	Antinematicidal action by inducing systemic resistance.	Medeiros et al. (2017)
12.	*Aspergillus flavus*	Insecticidal metabolites.	Senthilkumar et al. (2014)
13.	*Nigrospora* sp.	Antinematicidal activity.	Amin (2015)
14.	*Chaetomium globosum*	Antinematicidal activity.	Hu et al. (2013)
15.	*Botryosphaerica* sp.	Antinematicidal activity.	Chen et al. (2015)
16.	*Penicillium* sp.	Antifungal action towards plant pathogenic fungus.	Oliveira et al. (2009)
17.	*Trichoderma* sp.	Camptothecin production.	Pu et al. (2013)
18.	*Nigrospora* sp.	Antifungal action towards plant pathogenic fungus.	Wu et al. (2011)
19.	*Taxomyces andreanae*	Taxol production	Stierle et al. (1993)
20.	*Colletotrichum* sp.	Plant growth promoting activity.	Lu et al. (2000)
21.	*Muscodar albus*	Mycofumigant	Lacey et al. (2009)
22.	*Phoma* sp.	VOC active against phytopathogens.	Strobel (2011)
23.	*Microbotryum violaceum*	Antifungal action towards plant pathogenic fungus.	Hussain et al. (2009)
24.	*Fusarium solani*	Tolerance towards various biotic stress via SAR.	Kavroulakis et al. (2007)
25.	*Sarocladium strictum*	Pesticidal action against *Spodoptera littoralis*.	El-Sayed et al. (2020)

often fall quarry to such unsuitable practices while applying pesticides (Lee et al., 2004). Sustainable agriculture denotes preserving both magnitude and quality of agricultural produce for continued phases of time without unjustified resource enervation and vitality damage. Murphy et al. (2018) demonstrated an effective approach of delivering fungal endophyte consortia in barley crop in their longitudinal study. The principles of the said study can be adapted to any numbers of crops in a given environment (Figure 6.1).

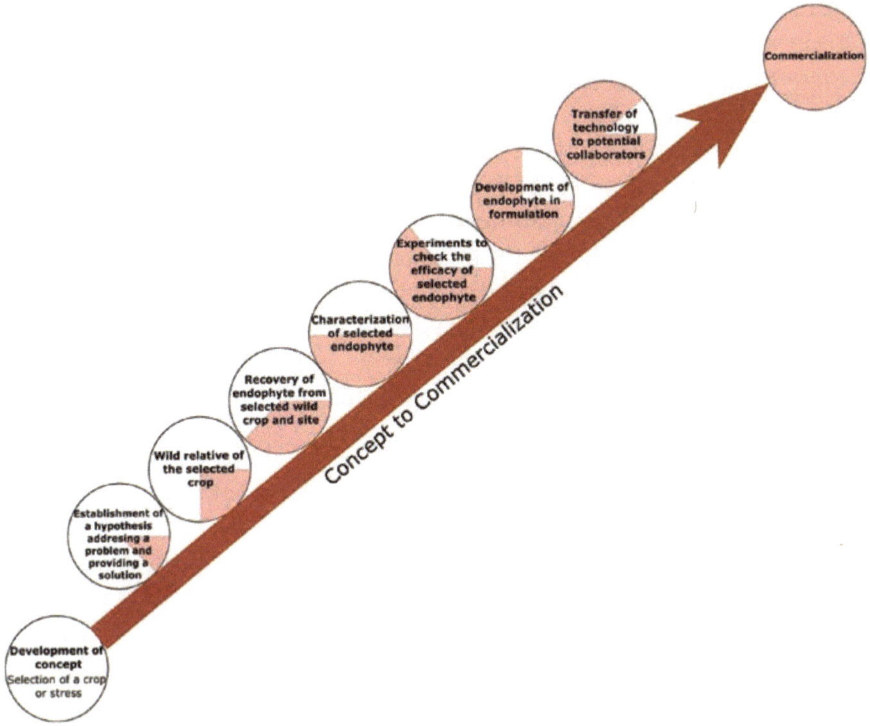

FIGURE 6.1 Development of concept to the commercialization of technology.

The overall microbial products required in the different industrial market is expected to reach > 250 billion USD in 2023, in which the pharmaceutical industry will have a share of about 40% with a yearly increment of >8% from 2017 onwards (Market Research Future, 2018). In agriculture, formulations based on *Bacillus* and *Pseudomonas* species are being recognized as highly potential (Sachdev & Cameotra, 2013). It is well-known fact that preserving the genetic diversity of animals and plants is vital for farm and food production. Till 2016, an estimated ~5 million samples of seeds and other plant's genetic material have been preserved in approximately 600 gene banks in international centers related to food and agriculture (https://sustainabledevelopment.un.org/sdg2). Additionally, Berg & Raaijmakers (2018) proposed that international seed banks should maintain seed-associated microbial diversity to save and study microbes which could be imperative for safeguarding global food production as well as economic capability.

The bottom-up process to examine biodiversity is rather exploratory and precisely correlates at the genetic and molecular level since lab screenings provide quite restricted information. As in a case, a particular *Pseudomonas* strain which established itself as antagonistic activity against *Phytophthora infestans*, then in the pot studies lost its biocontrol properties. In environmental conditions microbes become elusive to ward-off competition therefore, *in-vivo* studies are required to buttress the lab studies. Whereas in the case of top-down approaches it is cost-effective and practicable to directly identify the core and satellite microbiota in samples collected from natural habitats based on single amplicon variants in a high throughput sequencing analysis. The reward of the top-down approach is that its pre-selects the candidature microbes growing under field conditions being exposed to a multiple stress scenario while bottom-up screening approaches imitate field conditions in a controlled setup (Ke et al., 2021).

6.4 AGRICULTURAL PRACTICES AND MICROBIOME SHIFT

Agricultural practices have strongly impacted the plant microbiome system largely influenced by modified nutritional soil properties. Sustainable methods have been gaining ground at a worldwide level in agriculture and are identified as the most "dedicated" for preserving plant microbiomes. As per a recent study, usage of organic compost, manure, and agroforestry structure has strongly influenced the constitution of the bacterial microbiome, while its cultivation in monoculture fields favors abundant growth in the fungal microbiome. Lupatini et al. (2017) described that after amending organic compost in soil with crop rotation led to an increase in the phylogenetic richness, diversity, and microbial heterogeneity in comparison to the conventional cultivation system. For example, the addition of mulch in the potato rhizosphere, increased the diversity of fungi belonging to the Ascomycota phylum, simultaneously inhibiting the reproduction of the pathogenic strains of *Fusarium* (Santos & Olivares, 2021). As per an estimate, about 20% of the crop yield in the major food and cash crops is lost due to various plant diseases.

There are unquestionably many gaps in our information of plant-endophyte interactions, nonetheless, molecular operations such as genomics, proteomics, transcriptomics, and metabolomics are filling these gaps. Though, on-field (greenhouse) and pot experiments along with plate bioassays (*in-vitro*) are still essential for defining the functions of unidentified genes that are elaborated using genomics. Therefore, it is advocated to corroborate genomic and experimental studies together to support sustainable processes in diverse agricultural structures at the global level.

In a longitudinal comparative study involving *Cadophora* sp. and *Periconia macrospinosa* (dark septate fungi), and 32 non-endophytic fungi which categorically highlighted that endophytic strains contain multiple copies of genes encoding plant cell wall-degrading enzymes. These enzymatic activities depict that microbes indulging in an endophytic lifestyle have the ability to penetrate the tissues of the plant cell wall for initial colonization and to survive in such a micro-niche, have the ability to obtain nutrients (Knapp et al., 2018). Similarly, Fadiji et al. (2021) carried out the state-of-the-art functional diversity study of endophytic microbes inhabited in maize plants using shotgun metagenomics. The analysis revealed that the functional diversity of the microbiome does get impacted by different farming practices. The study indicated that traits related to plant growth promotion were highly signified in endophytes from plants cultivated in soil containing organic fertilizer. Further, the study recommends usage of organic fertilizers in maize cultivation and in promoting sustainable agriculture method (Figure 6.2).

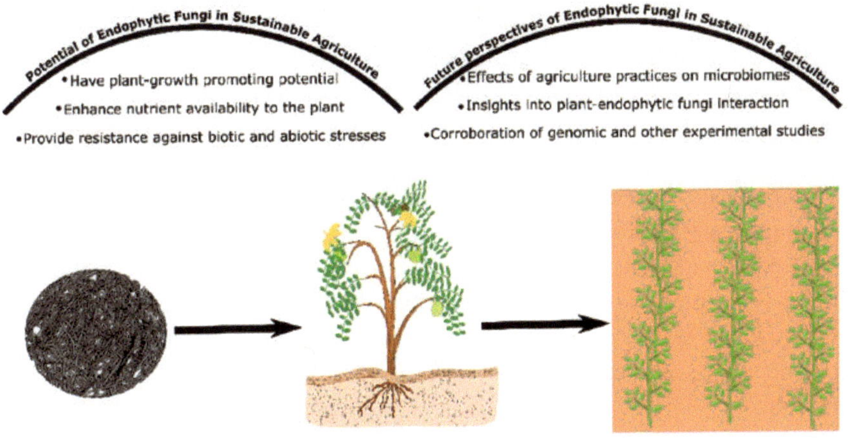

FIGURE 6.2 Potential and future perspectives of endophytic fungi in sustainable agriculture.

6.5 MICROBIOME ENGINEERING FOR SUSTAINABLE AGRICULTURE

Domestication of plants for food and feed has drastically altered soil physical and chemical structures and genetic diversity in plants. In order to meet the SDG goal of zero hunger index agriculturists are taking necessary steps to adopt region-wise appropriate sustainable strategies which cause a menial ecological footprint. To achieve sustainability, it is pertinent to make the best use of the surrounding natural processes which are functional in the arable land, for augmenting the plant's overall fitness and yield. The soil system not only acts as complex habitat but also is the reservoir of the microbiome. The microbes get attracted to plants via secretion of root exudates, initially to aid in nutrient recycling and exchange. This dynamic and resilient co-dependency of the plant and its associated microbiome gives researchers the opportunity to effectively manipulate and channel these interactions towards the enrichment of beneficial plant traits. Another aspect of exercising sustainability in agriculture is the plant genotypic characteristic which also strongly determines the structure and function of the plant-associated microbiome (Dubey et al., 2019; Dubey & Sharma, 2021).

6.6 UPDATED GENOMICS STUDIES RELATED TO ENDOPHYTES

The revelation of the molecular mechanism of the CRISPR/Cas system succeeded in a newer area of genome editing termed as RNA-guided engineered nucleases that utilize sequence-specific nucleases for highly detailed genetic modifications. With the aid of the CRISPR-RNP approach, two genetic clusters (ergot alkaloid biosynthesis genes; EAS1 and EAS2) producing toxic alkaloids were specifically deleted in seed-borne endophytic *Epichloë coenophiala* strain e19. The transformed endophytic fungus lacked the toxins and henceforth can be applied safely for forage cultivation and development.

Numerous studies have taken place recently to unravel molecular mechanisms relating to the endophytic-plant interactions and subsequently induced signal transduction. The various techniques/methods which have been formulated for characterizing specific functional domains within the life cycle of endophytes are: (i) differential genetic expression using the suppression subtractive hybridization (SSH) methodology; (ii) disease resistance mechanism was elucidated by sequencing oligonucleotide ligate and detecting-serial analysis of gene expression (SOLiD-SAGE); and (iii) hybridization-based microarrays for revealing endophytic genetic expression patterns, endophyte–endophyte and endophyte–plant interactions using mRNA deep sequencing (RNA-Seq) (Cosoveanu et al., 2021). For instance, high-throughput second-generation sequencing analysis describes the depiction of expression of reporter genes (GFP and DsRed) in *Epichloë* endophyte strains in a re-inoculation experimentation demonstrating the co-existence of endophytes in the same host suggesting the molecular basis of host-symbiont association (Hettiarachchige et al., 2019).

6.7 *PIRIFORMOSPORA INDICA*: MOST CELEBRATED ENDOPHYTE

Piriformospora indica is an axenically cultivable, root colonizing, biotrophic endophyte, belonging to the order Sebacinales of the basidiomycetes class (Azizi et al., 2021). *P. indica* has a wide range of host species ranging from vascular plant to mosses, which has been attributed to its ability to manipulate the metabolism of host plant while infection by delivering effector molecules in the host cell which eventually suppress the defense response, thus allowing and enhancing microbial infection (Gill et al., 2016). The development and colonization of *P. indica* in the host roots is a biphasic event, starting with the biotrophic phase using immune response suppression of host, followed by cell death phase, escaping any symptomatic developments (Gill et al., 2016). Due to its phytopromotional and phytoprotective capabilities, this endophyte has been considered to imitate the potentialities of arbuscular mycorrhiza.

The mutualistic association between *P. indica* and host root leads to improvement in several growth promotions and plant health indicators. Colonization of roots by *P. indica* has been observed to provide improved nutrient acquisition and enhanced biomass production in the host plant (Aslam et al., 2019). *P. indica* has the potential to mediate in Phosphorus (P) uptake and translocation in the host plant using phosphate transporter (PiPT), which otherwise is fixed in scarcely soluble form in soil (Aslam et al., 2019). Also, *P. indica* produces a phosphate liberating enzyme, acid phosphatase, which converts the P in the soil to plant accessible form. Wu et al. (2018) made the observation that under P deficient conditions, *P. indica* induces the PiPT activity, thus facilitating the accessibility of phosphorus present in the soil to the plant. *P. indica* also aids in the Nitrogen availability to the plant by activating the nitrate reductase enzyme that has an essential role in providing nitrate to the plant (Liu et al., 2020). Seed treated using *P. indica* have exhibited induced germination and better survival in adverse conditions (Mensah et al., 2020). Micro-propagated plantlets have also been observed to be immensely benefitted by *P. indica* application, which reduced its mortality after transplantation (Mensah et al., 2020). *P. indica* colonization in several plant species has shown stimulated production of SMs which can be harnessed for pharmaceutical, food, and agricultural purposes (Jisha et al., 2019).

P. indica has been widely known for its remediating potential against different abiotic and biotic stresses (Abo Nouh et al., 2020). In response to abiotic stresses such as heat, drought, salinity, etc., *P. indica* colonization in the host plant has been observed to establish and upregulate the ROS-scavenging system (Gill et al., 2016). This upregulated ROS-scavenging system also provides protection against biotic stresses such as viral, and fungal infections (Mensah et al., 2020). The constructive interaction between the host plant and *P. indica* confer systemically induced response in the plant against the pathogens involving the upregulation of pathogenesis-related, jasmonic acid (JA) and ethylene signaling genes (Khalid et al., 2019). Other than imparting the indirect induced resistance against the pathogen, *P. indica* has also exhibited direct antagonism against the invading pathogens (Gill et al., 2016). Hydroxamic acid, a secondary metabolite synthesized by *P. indica* has been known to have pesticidal potential (Gill et al., 2016). Consequently, with the incorporation of integrated farming and biotechnology-based approaches, *P. indica* can play a crucial role in crop improvement and sustainable agriculture.

ACKNOWLEDGEMENTS

This work was financially supported by DST-WOS B (DST/WOSB/2018/1858) given to KC. Authors clarify non-existence of any potential conflict of interest related with present research.

KEYWORDS

- **biofertilizer**
- **biopesticide**
- **CRISPR/Cas**
- **endophytes**
- **microbiome**
- **sustainable agriculture**

REFERENCES

Abo, N. F. A., Abo, N. H. H., & Abdel-Azeem, A. M., (2020). Piriformospora indica: Endophytic fungus for salt stress tolerance and disease resistance. *Agriculturally Important Fungi for Sustainable Agriculture*, 261–283.

Amin, N., (2015). Nematicidal activity of root exudates of Sengon plant inoculated with endophytic fungi *Nigrospora* sp. to control of root-knot nematode *Meloidogyne* spp. *J. Chem. Pharm. Res.*, *7*, 307–310.

Aslam, M. M., Karanja, J., & Bello, S. K., (2019). *Piriformospora indica* colonization reprograms plants to improved P-uptake, enhanced crop performance, and biotic/abiotic stress tolerance. *Physiological and Molecular Plant Pathology*, *106*, 232–237.

Azizi, M., Fard, E. M., & Ghabooli, M., (2021). Piriformospora indica affect drought tolerance by regulation of genes expression and some morphophysiological parameters in tomato (*Solanum lycopersicum* L.). *Scientia Horticulturae*, *287*, 110260.

Berg, G., & Raaijmakers, J. M., (2018). Saving seed microbiomes. *The ISME Journal*, *12*(5), 1167–1170.

Bücker, A., Bücker, N. C. F., Souza, A. Q. L. D., Gama, A. M. D., Rodrigues-Filho, E., Costa, F. M. D., Nunez, C. V., et al., (2013). Larvicidal effects of endophytic and basidiomycete fungus extracts on *Aedes* and *Anopheles* larvae (*Diptera, Culicidae*). *Revista Da Sociedade Brasileira de Medicina Tropical*, *46*, 411–419.

Chen, Y. M., Yang, Y. H., Li, X. N., Zou, C., & Zhao, P. J., (2015). Diterpenoids from the endophytic fungus *Botryosphaeria* sp. P483 of the Chinese herbal medicine *Huperzia serrata*. *Molecules*, *20*(9), 16924–16932.

Chowdhary, K., & Kaushik, N., (2019). Biodiversity and *in vitro* inhibition study of fungal endophytes of *Chlorophytum borivilianum* against selected phytopathogens. *Proceedings of the National Academy of Sciences, India Section B: Biological Sciences*, *89*(1), 113–121.

Chowdhary, K., & Kaushik, N., (2019). UPLC–MS and dereplication-based identification of metabolites in antifungal extracts of fungal endophytes. *Proceedings of the National Academy of Sciences, India Section B: Biological Sciences*, *89*(4), 1379–1387.

Chowdhary, K., & Sharma, S., (2020). Plant growth promotion and biocontrol potential of fungal endophytes in the inflorescence of *Aloe vera* L. *Proceedings of the National Academy of Sciences, India Section B: Biological Sciences*, *90*(5), 1045–1055.

Cosoveanu, A., Chowdhary, K., Cabrera, R., & Sharma, S., (2021). Role of phytohormones-producing fungal endophytes in plant–microbial interactions under stress. *Endophytes*, 195–223.

Costa, D., Tavares, R. M., Baptista, P., & Lino-Neto, T., (2020). Cork oak endophytic fungi as potential biocontrol agents against *Biscogniauxia mediterranea* and *Diplodia corticola*. *Journal of Fungi*, *6*(4), 287.

De Lamo, F. J., Šimkovicová, M., Fresno, D. H., De Groot, T., Tintor, N., Rep, M., & Takken, F. L. W., (2021). Pattern-triggered immunity restricts host colonization by endophytic fusaria, but does not affect endophyte-mediated resistance. *Molecular Plant Pathology*, *22*(2), 204–215.

Dubey, A., Malla, M. A., Khan, F., Chowdhary, K., Yadav, S., Kumar, A., Sharma, S., et al., (2019). Soil microbiome: A key player for conservation of soil health under changing climate. *Biodiversity and Conservation*, *28*(8), 2405–2429.

Dubey, S., & Sharma, S., (2021). Rhizospheric engineering by plant-mediated indirect selection of microbiome for agricultural sustainability. *Critical Reviews in Plant Sciences, 40*(5), 379–397.

El-Sayed, A. S. A., Abdel-Ghany, S. E., & Ali, G. S., (2017). Genome editing approaches: Manipulating of lovastatin and Taxol synthesis of filamentous fungi by CRISPR/Cas9 system. *Applied Microbiology and Biotechnology, 101*(10), 3953–3976.

El-Sayed, A. S. A., Moustafa, A. H., Hussein, H. A., El-Sheikh, A. A., El-Shafey, S. N., Fathy, N. A. M., & Enan, G. A., (2020). Potential insecticidal activity of *Sarocladium strictum*, an endophyte of *Cynanchum acutum*, against *Spodoptera littoralis*, a polyphagous insect pest. *Biocatalysis and Agricultural Biotechnology, 24*, 101524.

Fadiji, A. E., Ayangbenro, A. S., & Babalola, O. O., (2021). Shotgun metagenomics reveals the functional diversity of root-associated endophytic microbiomes in maize plant. *Current Plant Biology, 25*, 100195.

Gill, S. S., Gill, R., Trivedi, D. K., Anjum, N. A., Sharma, K. K., Ansari, M. W., Ansari, A. A., et al., (2016). *Piriformospora indica*: Potential and significance in plant stress tolerance. *Frontiers in Microbiology, 7*, 1–20. https://doi.org/10.3389/fmicb.2016.00332.

Hettiarachchige, I. K., Ludlow, E. J., Ekanayake, P. N., Brohier, N. D., Sahab, S., Sawbridge, T. I., Spangenberg, G. C., & Guthridge, K. M., (2019). Generation of epichloë strains expressing fluorescent proteins suitable for studying host-endophyte interactions and characterization of a T-DNA integration event. *Microorganisms, 8*(1), 54.

Hong, Z., Fang, S. S., Zhi, L. Z., & Zheng, F. M., (2010). Insecticidal activity of endophytic fungi from *Melia azedarach* L. *Chinese J. Biol. Control, 26*(1): 47-52.

Hu, Y., Zhang, W., Zhang, P., Ruan, W., & Zhu, X., (2013). Nematicidal activity of chaetoglobosin A poduced by *Chaetomium globosum* NK102 against *Meloidogyne incognita*. *Journal of Agricultural and Food Chemistry, 61*(1), 41–46.

Hussain, H., Krohn, K., Draeger, S., Meier, K., & Schulz, B., (2009). Bioactive chemical constituents of a sterile endophytic fungus from *Meliotus dentatus*. *Records of Natural Products, 3*(2).

Jisha, S., Kk, S., & Manjula, S. (2019). Multifunctional aspects of *Piriformospora indica* in plant endosymbiosis. *Mycology, 10*(3), 182. DOI: 10.1080/21501203.2019.1600063.

Kavroulakis, N., Ntougias, S., Zervakis, G. I., Ehaliotis, C., Haralampidis, K., & Papadopoulou, K. K., (2007). Role of ethylene in the protection of tomato plants against soil-borne fungal pathogens conferred by an endophytic *Fusarium solani* strain. *Journal of Experimental Botany, 58*(14), 3853–3864.

Ke, J., Wang, B., & Yoshikuni, Y., (2021). Microbiome engineering: Synthetic biology of plant-associated microbiomes in sustainable agriculture. *Trends in Biotechnology, 39*(3), 244–261.

Khalid, M., Rahman, S., & Huang, D., (2019). Molecular mechanism underlying *Piriformospora indica*-mediated plant improvement/protection for sustainable agriculture. *Acta Biochimica et Biophysica Sinica, 51*(3), 229–242.

Khruengsai, S., Pripdeevech, P., Tanapichatsakul, C., Srisuwannapa, C., D'Souza, P. E., & Panuwet, P., (2021). Antifungal properties of volatile organic compounds produced by *Daldinia eschscholtzii* MFLUCC 19-0493 isolated from *Barleria prionitis* leaves against *Colletotrichum acutatum* and its post-harvest infections on strawberry fruits. *PeerJ, 9*, e11242.

Knapp, D. G., Németh, J. B., Barry, K., Hainaut, M., Henrissat, B., Johnson, J., Kuo, A., et al., (2018). Comparative genomics provides insights into the lifestyle and reveals functional heterogeneity of dark septate endophytic fungi. *Scientific Reports*, *8*(1), 1–13.

Lacey, L. A., Horton, D. R., Jones, D. C., Headrick, H. L., & Neven, L. G., (2009). Efficacy of the biofumigant fungus *Muscodor albus* (*Ascomycota: Xylariales*) for control of codling moth (*Lepidoptera: Tortricidae*) in simulated storage conditions. *Journal of Economic Entomology*, *102*(1), 43–49.

Larkin, B. G., Hunt, L. S., & Ramsey, P. W., (2012). Foliar nutrients shape fungal endophyte communities in Western white pine (*Pinus monticola*) with implications for white-tailed deer herbivory. *Fungal Ecology*, *5*(2), 252–260.

Lee, W. J., Blair, A., Hoppin, J. A., Lubin, J. H., Rusiecki, J. A., Sandler, D. P., Dosemeci, M., & Alavanja, M. C. R., (2004). Cancer incidence among pesticide applicators exposed to chlorpyrifos in the agricultural health study. *Journal of the National Cancer Institute*, *96*(23), 1781–1789.

Liarzi, O., Bar, E., Lewinsohn, E., & Ezra, D., (2016). Use of the endophytic fungus *Daldinia* cf. concentrica and its volatiles as bio-control agents. *PLoS One*, *11*(12), e0168242.

Liu, Y., Jin-Li, C. A. O., Zou, Y. N., Qiang-Sheng, W. U., & Kamil, K., (2020). Piriformospora indica: A root endophytic fungus and its roles in plants. *Notulae Botanicae Horti Agrobotanici Cluj-Napoca*, *48*(1), 1–13.

Lu, H., Zou, W. X., Meng, J. C., Hu, J., & Tan, R. X., (2000). New bioactive metabolites produced by *Colletotrichum* sp., an endophytic fungus in *Artemisia annua. Plant Science*, *151*(1), 67–73.

Lupatini, M., Korthals, G. W., De Hollander, M., Janssens, T. K. S., & Kuramae, E. E., (2017). Soil microbiome is more heterogeneous in organic than in conventional farming system. *Frontiers in Microbiology*, *7*, 2064.

Medeiros, H. A. De, Araújo, F. J. V. De, Freitas, L. G. De, Castillo, P., Rubio, M. B., Hermosa, R., & Monte, E., (2017). Tomato progeny inherit resistance to the nematode *Meloidogyne javanica* linked to plant growth induced by the biocontrol fungus *Trichoderma atroviride*. *Scientific Reports*, *7*(1), 1–13.

Mensah, R. A., Li, D., Liu, F., Tian, N., Sun, X., Hao, X., Lai, Z., & Cheng, C., (2020). Versatile *Piriformospora indica* and its potential applications in horticultural crops. *Horticultural Plant Journal*, *6*(2), 111–121.

Mookherjee, A., Bera, P., Mitra, A., & Maiti, M. K., (2018). Characterization and synergistic effect of antifungal volatile organic compounds emitted by the *Geotrichum candidum* PF005, an endophytic fungus from the eggplant. *Microbial Ecology*, *75*(3), 647–661.

Murphy, B. R., Doohan, F. M., & Hodkinson, T. R., (2018). From concept to commerce: Developing a successful fungal endophyte inoculant for agricultural crops. *Journal of Fungi*, *4*(1), 24.

Oliveira, C. M., Silva, G. H., Regasini, L. O., Zanardi, L. M., Evangelista, A. H., Young, M. C. M., Bolzani, V. S., & Araujo, A. R., (2009). Bioactive metabolites produced by *Penicillium* sp. 1 and sp. 2, two endophytes associated with *Alibertia macrophylla* (*Rubiaceae*). *Zeitschrift Für Naturforschung C*, *64*(11, 12), 824–830.

Pu, X., Qu, X., Chen, F., Bao, J., Zhang, G., & Luo, Y., (2013). Camptothecin-producing endophytic fungus *Trichoderma atroviride* LY357: Isolation, identification, and fermentation conditions optimization for camptothecin production. *Applied Microbiology and Biotechnology*, *97*(21), 9365–9375.

Rosa, L. H., Tabanca, N., Techen, N., Wedge, D. E., Pan, Z., Bernier, U. R., Becnel, J. J., et al., (2012). *Diversity and Biological Activities of Endophytic Fungi Associated with Micropropagated Medicinal Plant Echinacea Purpurea (L.) Moench*. Agricultural Research Service Gainesville Fl Center for Medical Agricultural.

Sachdev, D. P., & Cameotra, S. S., (2013). Biosurfactants in agriculture. *Applied Microbiology and Biotechnology*, *97*(3), 1005–1016.

Sánchez-Fernández, R. E., Diaz, D., Duarte, G., Lappe-Oliveras, P., Sánchez, S., & Macías-Rubalcava, M. L., (2016). Antifungal volatile organic compounds from the endophyte *Nodulisporium* sp. strain GS4d2II1a: A qualitative change in the intraspecific and interspecific interactions with *Pythium aphanidermatum*. *Microbial Ecology*, *71*(2), 347–364.

Santos, L. F., & Olivares, F. L., (2021). Plant microbiome structure and benefits for sustainable agriculture. *Current Plant Biology*, *26*, 100198.

Senthilkumar, N., Murugesan, S., & Babu, D. S., (2014). Metabolite profiling of the extracts of endophytic fungi of entomopathogenic significance, *Aspergillus flavus* and *Nigrospora sphaerica* isolated from tropical tree species of India, *Tectona grandis* L. *Journal of Agriculture and Life Sciences, 1*(1).

Stierle, A., Strobel, G., & Stierle, D., (1993). Taxol and taxane production by *Taxomyces andreanae*, an endophytic fungus of Pacific yew. *Science*, *260*(5105), 214–216.

Strobel, G., (2011). Muscodor species-endophytes with biological promise. *Phytochemistry Reviews*, *10*(2), 165–172.

Wu, M., Wei, Q., Xu, L., Li, H., Oelmüller, R., & Zhang, W., (2018). *Piriformospora indica* enhances phosphorus absorption by stimulating acid phosphatase activities and organic acid accumulation in *Brassica napus*. *Plant and Soil*, *432*(1), 333–344.

Wu, S. H., Chen, Y. W., & Miao, C. P., (2011). Secondary metabolites of endophytic fungus *Xylaria* sp. YC-10 of *Azadirachta indica*. *Chemistry of Natural Compounds*, *47*(5), 858–861.

CHAPTER 7

Impact of Abiotic and Biotic Stresses on Agricultural Crop

MEENA SINGH, AKANKSHA KUMARI, and POONAM DEHARIYA

Lab of Mushroom Biology, Department of Botany,
Dr. Harisingh Gour University, Sagar, Madhya Pradesh, India

ABSTRACT

Higher plants are sessile, which means they can't escape abiotic stress. They are constantly being exposed to different abiotic stressors with no safeguards in place. Animals, on the other hand, can move around and run away from danger. Plants need to be protected more because of their stationary nature. Plants tend to adapt to harsh environmental changes throughout evolution. They've adapted to various ecological niches and survived. A range of abiotic and biotic challenges are brought together by global warming and reducing agricultural production. Scientists have been working in physiology, molecular biology, and genetics to better understand crop response to these stresses and identify mechanisms that regulate plant adaptation and tolerance to stress in recent years. This chapter gives a thorough overview of various abiotic and biotic stresses and their effects on crops, which will aid physiologists and molecular biologists in developing agronomically appropriate methods for the development of broad-spectrum stress-resistant crops.

7.1 INTRODUCTION

Plant stress is defined as "a state in which increasing demands placed on a plant cause an initial destabilization of functions, followed by normalization and improved resistance," as well as "if the limits of tolerance are exceeded

Advances in Plant Microbiome Research for Climate Resilient Agriculture: Toward Sustainable Farming.
Ashwani Kumar (Ed.)

and the adaptive capability is overloaded, the outcome could be severe harm or even death," (Larcher, 1987). The statement that follows the fundamental general stress principle for living things: "All agents can function as stressors, creating both stress and specific action," and "There exist stressor-specific responses and non-specific general responses" (Hans Selye, 1936). Lichtenthaler (1987, 1988) broadened the idea of plant stress by including the phase of regeneration, when stresses are eliminated, and recognizing between eu-stress and dis-stress. Dis-stress (as seen in the English word distress) is significant stress that negatively affects the plant and causes harm. Eu-stress is activating, stimulating stress, and a good ingredient for plant development. "A little stress can activate cell metabolism and improve a plant's physiological activity while causing no long-term harm." The plant benefits from such modest stimulating stress.

Mittler and his colleagues created a "stress matrix" to organize the effects of numerous abiotic and biotic stresses on plant development and efficiency (Suzuki et al., 2014; Mittler, 2006). Matrix shows that stress combinations can have both positive and negative impacts on plants. Consequently, figuring physio-morphological characteristics that are affected by combined pressures is necessary for the development of some plants that have a greater tolerance for combined abiotic and biotic challenges. Abiotic stresses include heat and drought that occur simultaneously, crop productivity is more negatively impacted than when these conditions are experienced separately at various periods of crop growth (Mittler, 2006; Prasad et al., 2011). Drought, high and low temperatures, and salinity have all been shown to influence the prevalence and spread of diseases, weeds, and insects (Coakley et al., 1999). They may also cause minor pests to develop into future dangers (Duveiller et al., 2007). Plants only choice for dealing with stress is to change their physiologies, metabolic systems, gene expressions, and developmental activities. Individual, multiple individuals, and combined pressures are the three types of stress. Single stress is the impact of only one stress factor on the growth and development of plant, multiple stress, in contrast, is the result of more than two stresses occurrence during different times without overlap (multiple individuals) or concurrently with at least some overlap (multiple individuals) (combined). The combined abiotic stress is when drought and heat stress occur simultaneously in the summer, whereas the combined biotic stress is when a bacterial and fungal pathogen attacks a plant at the same time. For example, *Cladosporium* spp., *Alternaria* spp., *Colletotrichum* spp., *Phomopsis* spp., and *Fusarium* spp., as well as a bacterium, *Xanthomonas arboricola* induce brown apical necrosis in *Juglans regia* (walnut) (Belisario et al., 2002).

Plant stress is described as a negative impact on a plant's physiology brought on by a quick change from an ideal environmental condition in which homeostasis is preserved to an unfavorable environment in which this initial homeostatic state is broken. Therefore, the word "plant stress" is a relative one because it is always used to compare the physiological effects of a stress-induced physiological phenomenon in a plant species to the physiological effects of the same physiological phenomenon in the same plant species under optimal conditions. Plants had evolved a variety of mechanisms to accommodate changes in their fluctuating growth conditions, enabling flexibility in function under the effect of environmental factors with no disrupting cellular and developmental biological processes by creating a number of secondary metabolites (SMs) which perform a wide variety of roles in responses to shifting environmental conditions., growth, and development. Plant's ability to produce metabolites is regarded as an adaptive capacity for coping with stressful constraints in a challenging and changing growth environment, which may include the formation of intricate chemical forms and interactions in structure and functional stabilization via routes and signaling processes (TasiuIsah, 2019). Agriculture faces the enormous problem of feeding a growing global population while maintaining natural resource availability. Agronomy and associated sciences will undoubtedly support this big task in the future (Gimenez et al., 2018).

Abiotic stressors can change plant–pest interactions by increasing host plant vulnerability to pathogenic organisms, insects, and weeds, as well as lowering competitive capacity. Some pests, on the other hand, may change plant responses to abiotic stress factors (Pandey et al., 2017). Heat stress combined with salt stress is a common occurrence for plants in dry and semi-arid environments. Heat stress is frequently accompanied by high light stress. *Vitis vinifera* (grapes) grown in continental climates, such as North China, are subjected to a combination of cold stress and drought that decreased their productivity (Su et al., 2015). Plants in the Mediterranean have a combination of cold and high light stress (Loreto & Bongi, 1989). Winter wheat, *Triticum aestivum*, is also well known to experience cold stress brought on by ozone, which reduces its ability to withstand frost (Barnes & Davison, 1988). Similarly, ozone stress mixed with salt decreases yields of *Oryza sativa* (rice) and *Cicer arietinum* (chickpea) (Welfare et al., 2002). There are two main categories of plant stress – biotic and abiotic stresses.

Abiotic stress includes, deficient or excessive water, low or high temperature, heavy metals, high salinity, oxidative stress, UV radiation and mechanical stress (Figure 7.1).

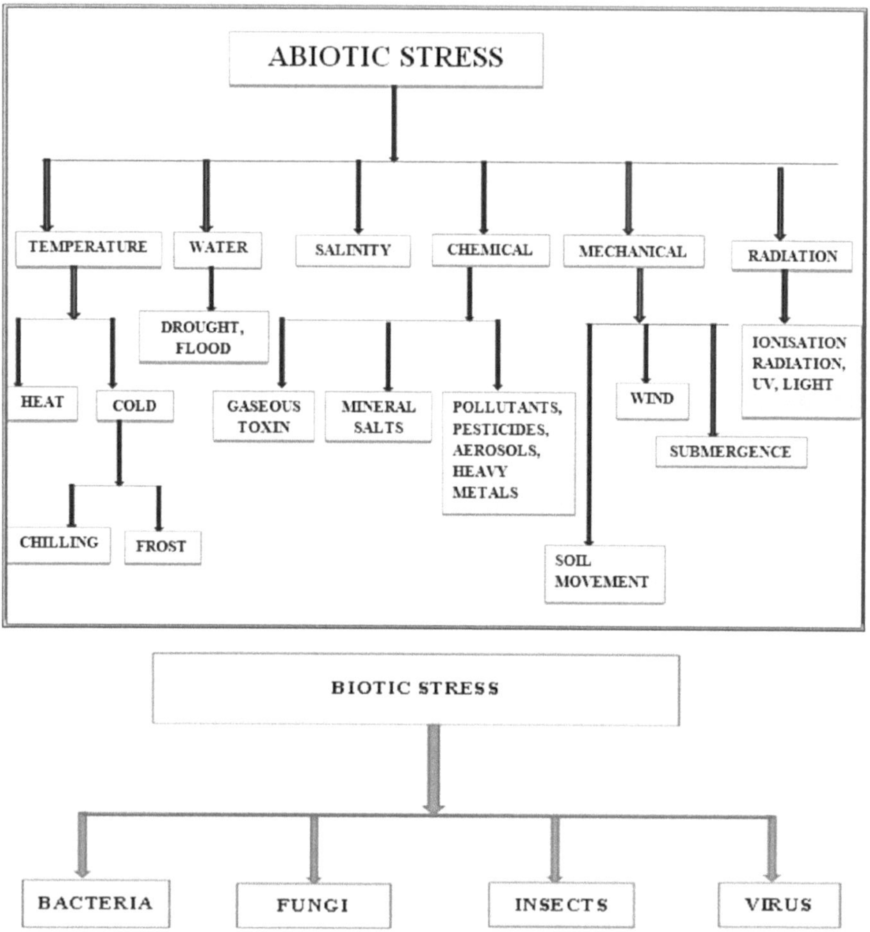

FIGURE 7.1 A flow chart of various biotic and abiotic stressors is represented diagrammatically.
Source: Adapted from Chhaya et al. (2021).

7.2 DEFICIENT OR EXCESSIVE WATER

Drought enhances competitive interactions between weeds and crops, as certain weeds have higher water usage more effective than crops (Valerio et al., 2013; Ziska et al., 2010; Patterson, 1995). Drought stress had a favorable transgenerational influence on the vigor of *Brassica napus* seedlings, according to Hatzig et al. (2018). They put this down to heterotic effects, a changed a source of metabolites for seed storage and a stress memory that spans generations created by stress-induced changes in the seedling's epigenome. Drought stress causes a significant reduction in the growth and yield of crops like *Hordeum vulgare*, *Oryza sativa*, *Zea mays* and *Triticum aestivum* (Lafitte et al., 2007; Rampino et al., 2006; Samarah, 2005; Kamara et al., 2003), due to low plant humidity and intense sunshine, and high temperature caused by drought with increased respiration, photosynthesis (Fathi & Tari, 2016).

Water stress causes development to slow down, the stomatal aperture to contract, and nutrient deficiencies (such as K^+ and Ca^{2+}) to occur (Chaves et al., 2009). When plants are subjected to long-term drought stress, root extension occurs, which could be explained by the plant's need to find groundwater (Brunner et al., 2015).

7.3 LOW OR HIGH TEMPERATURE

Heat stress has known effects on organisms life processes, either directly or through the change of surrounding environmental components and is expected to climb by 0.2°C every decade by 2100, resulting in temperatures 1.8–4.0°C warmer than the present temperature. Oxidative stress is brought on by an excessive generation of reactive oxygen species (ROS), is one of the principal effects of high-temperature stress. Oxidative stress, seed inhibition, reduced plant growth, incorrect development, altered photosynthesis, altered phenology, altered dry matter partitioning, water loss, yield reduction, and replication of crop quality are the main effects of high temperatures on plants. Heat stress has an impact on all elements of plant functioning, including germination, growth, development, reproduction, yield, and the stability of different proteins, RNA species, membranes, and cytoskeleton components, as well as altering the efficiency of enzymatic reactions in the cell, creating metabolic imbalance. One of the most heat-sensitive physiological processes in plants is photosynthesis. The principal sites of damage at HTs in chloroplast are carbon metabolism in the stroma and photochemical processes in thylakoid lamellae. Under heat stress, the thylakoid membrane is very vulnerable to HT, and considerable changes in chloroplasts occur, such as changed reduction in grana stacking, thylakoid structural organization, grana swelling, and activity of photosystem II (PSII) is drastically decreased or even ceases. Heat stress causes oxidation of chloroplast lipids and the thylakoid membranes, which affects leaf water status, stomatal conductance of leaves and level of intracellular CO_2. Another explanation for reduced photosynthesis in HT is stomata closing., which impacts intercellular CO_2. Heat stress for a short duration of time during reproduction can produce considerable reductions in floral buds and flower abortion, notwithstanding the significant differences in sensitivity between and within plant species and various types (Mirza et al., 2013).

Chilling or low-temperature stress impacts a variety of physiological and biochemical processes in plants, including cellular cytoskeletal structure and function, nucleic acid and protein synthesis, water, and nutrition cycle imbalances, and photosynthetic and respiratory metabolism inhibition. Primary and secondary events have been used to divide chilling injury into two types based on direct and indirect consequences. The primary symptom triggers the membrane phase change from liquid to solid gel, which reduces membrane permeability, or fluidity, and increases secondary symptom expression. Although enzymes are more labile when exposed to high temperatures, low temperatures have an impact on their efficiency. Several metabolic enzymes, including pyruvate P1 dikinase and phosphofructokinase, which play essential roles in C4 plant carbon fixation and glycolysis, are more influenced by freezing temperature because it changes the structure of the enzyme. Low temperature also affects K-mediated ATPase, as inhibited action causes K ion leakage from cells. Chilling stress affects enzyme kinetics because it reduces the enzyme's reaction velocity (Vmax) and affinity (Km) for its substrate (Hussain et al., 2018).

7.4 HEAVY METALS

Based on their toxicity, frequency of occurrence, and most importantly their potential for exposure to flora and fauna, the Agency for term Toxic Substances and Disease Registry (ATSDR, 2003) has determined that As, Pb, Cd, and Hg are the four heavy metals that are the most dangerous. Inorganic arsenate As(V) and arsenite As(III) are the two kinds of arsenate that are present in the environment. Both As(III) and As(V) are harmful and are categorized as important environmental contaminants by the United States Environmental Protection Agency (USEPA) (Tripathi et al., 2007; Verma et al., 2017). In plants, oxidative phosphorylation and the production of ATP are inhibited by As(V) throughout the energy metabolism process (Verma et al., 2016; Carbonell et al., 1998).

One of the most widely and consistently distributed trace elements in natural sources is lead. Leaded fuel contamination, substance, old lead plumbing lines, different locations for industry, or even old the orchard sites used in lead arsenate manufacturing might harm the health of the soil, flora, and wildlife (Tangahu et al., 2011). Pb^{2+} is non-biodegradable, and Acute toxicity results from long-term exposure to plants and animals. According to Bradham et al., (2006); and Pehlivan et al., (2009), it also negatively affects a number of biological processes, including soil characteristics like pH, organic carbon, amorphous iron, cation exchange capacity and aluminum oxides (FEAL). Pb toxicity causes ATP inhibition, lipid peroxidation, and DNA damage due to an excess of ROS. If appropriate remedial action is ignored, excessive soil Pb levels could never go back to normal (Traunfeld & Clement, 2001). One of the heavy metals that affects plants the most is cadmium. Because it is extremely soluble in water, plants may absorb it with ease, making it the principal entrance conduit into the food chain, posing substantial human health risks (Buchet et al., 1990). Cd is classified as a very serious human carcinogen by the International Agency for Research on Cancer (IARC, 1993).

Mercury is a naturally occurring component of the crust of the earth that builds up on land and aquatic ecosystems, primarily as a result of human activities such as mining and manufacturing (Järup, 2003). Low amounts of Hg^{2+} may not have an impact on plant growth, but at greater amounts, it becomes extremely phytotoxic to plant cells, causing obvious lesions and physiological issues (Zhou et al., 2007; Ortega-Villasante et al., 2005). Hg^{2+} binding to water channel proteins causes leaf stomata to close and water flow to be physically obstructed in plants (Zhang & Tyerman, 1999).

7.5 HIGH SALINITY

Salinity can also improve plant fertility by increasing clonal/sexual reproduction at certain concentrations (Van Zandt et al., 2003). Plants suffer from ionic stress as a result of long-term salt exposure, which causes leaf senescence and photosynthetic impairment, which has an extra detrimental impact on growth (Chaves et al., 2009). Heavy roots may acquire higher levels of chloride during long-term salinity stress. Excessive ion accumulation, particularly Na^{2+}, has a deleterious impact on photosynthetic components, lowering enzyme activity and pigment synthesis. The excess light received that is not utilized by the plant could rise ROS generation and as a result, oxidative stress. Many species have defenses against salt intrusion that keep it out of plant cells, or they can compartmentalize salt in the vacuoles to reduce salt content in the cytoplasm (Khan et al., 2019).

In salinity acclimated plants, some major metabolic byproducts, such as amino acids, carbohydrates, and nitrogen, play a crucial protective role by acting as osmolytes and ROS scavengers and shielding membranes and protein (Sharma et al., 2019).

7.6 OXIDATIVE STRESS

Aerobic metabolism inevitably contains ROS. Electron transport mechanisms in various subcellular compartments, such as the chloroplasts, cytosol, microbodies, and mitochondria, sequentially reduce molecular oxygen to create $^1O_2$ singlet oxygen, H_2O_2 hydrogen peroxide, O_2^- superoxide radicals, and OH hydroxyl radical (Dias et al., 2014). At lower concentrations, they play crucial signaling functions in a variety of processes, including seed germination, pollen tube development, senescence, root hair growth, leaf development, embryo formation, cell elongation, gravitropism, self-incompatibility, and many others (Gechev et al., 2006). When subjected to oxidative, abiotic, wounding, or pathogen stressors, ROS are either synthesized or detoxified, which causes them to activate and/or control the primary and secondary signaling pathways (Duque et al., 2013). According to Sharma et al., (2012), salicylic acid (SA) is also said to control ROS production when a wound is present.

A significant change in a redox environment is always brought about by the excessive generation of ROS that stress situations produce. Due to the disruption of signaling pathways, free radicals and other intermediate chemicals are scavenged and detoxified via antioxidant systems (Yun et al., 2010). Surplus of O_2^- activates superoxide dismutase (SOD) which converts O_2^- to H_2O_2 and the latter is removed by ascorbate peroxidase (APX). As a result, during diverse abiotic stresses, these antioxidant systems remove excess ROS that are not needed for fundamental plant functions and stabilize the fundamental biochemical state of the cell, promoting tolerance as well as acclimatization (Scandalios, 2005; Bowler & Fluhr, 2000).

By reducing the amount of water accessible to the plant, salinity stress generates physiological drought-like conditions and leads to ion imbalances. Under these circumstances, APX gives the affected plants varying degrees of salt tolerance. While constitutive over-expression lines exhibit enhanced resistance to 100 mM NaCl stress, the cytosolic APX (cAPX) mutant with lost activity exhibits vulnerability to salinity-induced oxidative stress (Diaz-Vivancos et al., 2013). In order to make up for the loss of APX and to provide stress tolerance, APX deficient mutants can upregulate other peroxidases (POD) (Narendra et al., 2006).

7.7 UV RADIATION

UV-B causes physiological damage to the photosynthetic machinery, DNA damage, membrane impacts, and morphological damage such as plant stunting and leaf discoloration, as well as a reduction in vegetative biomass and grain yield in plants. Extensive research on a variety of crops, including wheat, maize, cucumber, rice, soybean, and others, has revealed significant interspecific heterogeneity in plant responses to UV-B radiation, however, the cause for these differences has yet to be thoroughly investigated. Extensive studies on a variety of crops, including wheat, maize, cucumber, rice, soybean, and others, have revealed significant interspecific heterogeneity in plant responses to UV-B radiation, however, the cause for these differences has yet to be thoroughly investigated (Sharma et al., 2017).

UVB radiation also affects the reproductive or floral morphology of crop plants, resulting in yield reduction. Cotton flowers produced on plants exposed to UVB treatments were found to be smaller, with smaller petal and bract sizes and fewer anthers (Kakani et al., 2003a, b).

UV radiation has a greater impact in tropical climes because plants are exposed to sunshine for long periods. Studies on tropical crops such as *Vigna mungo*, *Vigna radiata*, and *Glycine max* (Mazza et al., 2000; Amudha et al., 2005; Guruprasad et al., 2007), *Triticum aestivum* (Kataria & Guruprasad, 2014), *Amaranthus* tricolor variants (Kataria & Guruprasad, 2014), and *Oryza sativa* (Teramura et al., 1991) when exposed to ambient UVB, development is slowed, leaf area expansion is reduced, and yield is reduced. Direct DNA damage is the most likely cause of the decrease in growth (Giordano et al., 2004).

7.8 MECHANICAL STRESS

Recent research shows that mechanical stimulation caused by wind used on the beans reduced spider mites egg formation and development on foliar disc of mechanically stimulated crops as well as on whole plants. *Trichoplusia ni* larvae in Arabidopsis were shown to be lighter when touched. Touch was recently discovered to reduce the susceptibility of maize and bean plants to aphids, in addition to its efficient effect on herbivores. Wounding MS has been shown to cause the production of ROS in certain investigations. Non-wounding stimuli also caused the creation of ROS and that the quantity of the ROS signal appears to be correlated with the number of rubs.

ROS production is essential for lowering tomato sensitivity to *B. cinerea*. According to a study on the impact of MS on fruit quality, MS promotes the development of antioxidant molecules (Coutand, 2020).

7.9 CONCLUSION

Stress factors that are both biotic and abiotic can alter metabolism and physiology changes in plants. There is a constant rise in temperature, inconsistent rainfall, and fluctuations in drought and flood are always taken into account. Salt stress can be greatly exacerbated by anthropogenic activities such overuse of fertilizers, poor irrigation practices, and resource exploitation. Plants will likely experience both biotic and abiotic stressors more frequently under these conditions. To ensure both the safety of the farmers and the security of the food supply, it is the responsibility of plant breeders to create cultivars that are able to endure stress. Growth of the plant may be impacted either badly or favorably by the relationship between the two stress conditions. For instance, a co-existing drought can also modify how certain diseases and plants interact, either suppressing or enhancing pathogen growth. In order to better understand the overall effect of stress combinations on plants, it is crucial to investigate how the two stresses interact. The main causes of abiotic and biotic stress in agricultural crops under various climatic situations are covered in this chapter. In developing arid and semiarid regions of the tropics, abiotic stresses brought on by temperature, water, salinity, chemical, mechanical, and radiation pose significant risks to food security. When it comes to crop failure and yield loss in the tropics, water and temperature stresses have been the most economically significant to crop output. There is a need to develop adaptation strategies for abiotic stress tolerance in agriculture so as to sustain crop production all year long regardless of variability in weather conditions. This is due to the numerous effects of abiotic stress as well as unfavorable projected changes in future climatic conditions. In order to maintain future crop production predictions, research on environmental stress on agricultural crops should be given attention. This is because there is a conflict between the increased need for food and the increased pressure on the environment that comes from an expanding human population, particularly in tropical countries of Africa and Asia.

KEYWORDS

- **abiotic stresses**
- **adaptation**
- **heavy metals**
- **oxidative stress**
- **physiology**
- **tolerance**
- **UV radiation**

REFERENCES

Amudha, P., Jayakumar, M., & Kulandaivelu, G., (2005). Impacts of ambient solar UV (280–400 nm) radiation on three tropical legumes. *Journal of Plant Biology, 48*, 284–291.

Barnes, J. D., & Davison, A. W., (1988). The influence of ozone on the winter hardiness of Norway Spruce [*Piceaabies* (L) Karst]. *New Phytol., 108*, 159–166. doi: 10.1111/j.1469-8137.1988.tb03692.x.

Belisario, A., Maccaroni, M., Corazza, L., Balmas, V., & Valier, A., (2002). Occurrence and etiology of brown apical necrosis on Persian (English) walnut fruit. *Plant Dis., 86*, 599–602. doi: 10.1094/PHYTO-01-15- 0029-R.

Bowler, C., & Fluhr, R., (2000). The role of calcium and activated oxygens as signals for controlling cross-tolerance. *Trends Plant Sci., 5*, 241–246. doi: 10.1016/S1360-1385(00)01628-9.

Bradham, K. D., Dayton, E. A., Basta, N. T., Schroder, J., Payton, M., & Lanno, R. P., (2006). Effect of soil properties on lead bioavailability and toxicity to earthworms. *Environ. Toxicol. Chem., 25*, 769–775. doi: 10.1897/04-552R.1.

Brunner, I., Herzog, C., Dawes, M. A., Arend, M., & Sperisen, C., (2015). How tree roots respond to drought. *Front. Plant Sci., 6*, 547. doi: 10.3389/fpls.2015.00547.

Buchet, J. P., Lauwerys, R., Roels, H., Bernard, A., Bruaux, P., Claeys, F., et al., (1990). Renal effects of cadmium body burden of the general population. *Lancet, 336*, 699–702. doi: 10.1016/0140-6736(90)92201-R.

Carbonell, A. A., Aarabi, M. A., Delaune, R. D., Grambrell, R. P., & Patrick, W. H., (1998). Arsenic in wetland vegetation: Availability, phytotoxicity, uptake and effects on plants growth and nutrition. Sci. *Total Environ., 217*, 189–199. doi: 10.1016/S0048-9697(98)00195-8.

Chaves, M. M., Flexas, J., & Pinheiro, C., (2009). Photosynthesis under drought and salt stress: Regulation mechanisms from whole plant to cell. *Ann. Bot., 103*, 551–560. doi: 10.1093/aob/mcn125.

Chhaya, Y. B., Jogawat, A., Gnanasekaran, P., Kumari, P., Lakra, N., Lal, S. K., & Narayan, O. P., (2021). An overview of recent advancement in phytohormones-mediated stress management and drought tolerance in crop plants. *Plant Gene, 25*, 100264.

Coakley, S. M., Scherm, H., & Chakraborty, S., (1999). Climate change and plant disease management. *Annu. Rev. Phytopathol.*, *37*, 399–426. doi: 10.1146/ annurev.phyto.37.1.399.

Coutand, C., (2020). The effect of mechanical stress on plant susceptibility to pests: A mini opinion review. *Plants*, *9*, 632. doi: 10.3390/plants9050632.

Dias, M. C., Azevedo, C., Costa, M., Pinto, G., & Santos, C., (2014). Meliaazedarach plants show tolerance properties to water shortage treatment: An ecophysiological study. *Plant Physiol. Biochem.*, *75*, 123–127. doi: 10.1016/j.plaphy.2013.12.014.

Diaz-Vivancos, P., Faize, M., Barba-Espin, G., Faize, L., Petri, C., Hernández, J. A., et al., (2013). Ectopic expression of cytosolic superoxide dismutase and ascorbate peroxidase leads to salt stress tolerance in transgenic plums. *Plant Biotech. J.*, *11*, 976–985. doi: 10.1111/pbi.12090.

Duveiller, E., Singh, R. P., & Nicol, J. M., (2007). The challenges of maintaining wheat productivity: Pests, diseases, and potential epidemics. *Euphytica*, *157*, 417–430. doi: 10.1007/s10681-007-9380-z.

Fathi, A., & Tari, D. B., (2016). Effect of drought stress and its mechanism in plants. *Int. J. Life Sci.*, *10*, 1–6. doi: 10.3126/ijls.v10i1.14509.

Gechev, T. S., Van, B. F., Stone, J. M., Denev, I., & Laloi, C., (2006). Reactive oxygen species as signals that modulate plant stress responses and programmed cell death. *Bioessays*, *28*, 1091–1101. doi: 10.1002/bies.20493.

Gimenez, E., Salinas, M., & Agugliaro, F. M., (2018). Worldwide research on plant defense against biotic stresses as improvement for sustainable agriculture. *Sustainability*, *10*, 391. doi: 10.3390/su10020391.

Giordano, C. V., Galatro, A., Puntarulo, S., & Ballaré, C. L., (2004). The inhibitory effects of UV-B radiation (280–315 nm) on *Gunnera magellanica* growth correlate with increased DNA damage but not with oxidative damage to lipids. *Plant, Cell & Environment*, *27*, 1415–1423.

Guruprasad, K., Bhattacharjee, S., Kataria, S., Yadav, S., Tiwari, A., Baroniya, S., Rajiv, A., & Mohanty, P., (2007). Growth enhancement of soybean (*Glycine max*) upon exclusion of UV-B and UV-A components of solar radiation, characterization of photosynthetic parameters in leaves. *Photosynthesis Research*, *94*, 299–306.

Hasanuzzaman, M., Nahar, K., Alam, M. M., Chowdhury, R. R., & Fujita, M., (2013). Physiological, biochemical, and molecular mechanisms of heat stress tolerance in plants. *Int. J. Mol. Sci.*, *14*, 9643–9684. doi: 10.3390/ijms14059643.

Hatzig, S. V., Nuppenau, J. N., Snowdon, R. J., & Schiessl, S. V., (2018). Drought stress has transgenerational effects on seeds and seedlings in winter oilseed rape (*Brassica napus* L.). *BMC Plant Biol.*, *18*, 297. doi: 10.1186/s12870-018-1531-y.

Hussain, H. A., Hussain, S., Khaliq, A., Ashraf, U., Anjum, S. A., Men, S., et al., (2018). Chilling and drought stresses in plants: Implication, cross talk, and potential management opportunities. *Front. Plant Sci.*, *9*, 393. Available from: https://doi.org/10.3389/fpls.2018.00393.

IARC, (1993). IARC monographs on the evaluation of carcinogenic risk to humans. *Beryllium, Cadmium, Mercury and Exposures in the Glass Manufacturing Industry* (Vol. 58, pp. 41–117). Lyon: International Agency for Research on Cancer Publication.

Järup, L., (2003). Hazards of heavy metal contamination. *Br. Med. Bull.*, *68*, 167–182. doi: 10.1093/bmb/ldg032.

Kakani, V. G., Reddy, K. R., Zhao, D., & Mohammed, A. R., (2003). Ultraviolet-B radiation effects on cotton (*Gossypium hirsutum* L.) morphology and anatomy. *Annals of Botany, 91*(7), 817–826.

Kakani, V. G., Reddy, K. R., Zhao, D., & Sailaja, K., (2003). Field crop responses to ultraviolet B radiation: A review. *Agricultural and Forest Meteorology, 120*, 191–218.

Kamara, A. Y., Menkir, A., Badu-Apraku, B., & Ibikunle, O., (2003). The influence of drought stress on growth, yield, and yield components of selected maize genotypes. *J. Agric. Sci., 141*, 43–50. doi: 10.1017/S0021859603003423.

Kataria, S., & Guruprasad, K. N., (2014). Exclusion of solar UV components improves growth and performance of *Amaranthus tricolor* varieties. *Scientia Horticulturae, 174*, 36–45.

Khan, A. L., Muneer, S., Kim, Y., Al-Rawahi, A., & Al-Harrasi, A., (2019). Silicon and salinity: Crosstalk in crop-mediated stress tolerance mechanisms. *Front. Plant Sci., 10*, 1429. doi: 10.3389/fpls.2019.01429.

Lafitte, H. R., Yongsheng, G., Yan, S., & Li, Z. K., (2007). Whole plant responses, key processes, and adaptation to drought stress: The case of rice. *J. Exp. Bot., 58*, 169–175. doi: 10.1093/jxb/erl101.

Larcher, W., (1987). Streβ beipflanzen, *Naturwissenschaften, 74*, 158167.

Lichtenthaler, H. K., (1988). *In vivo* chlorophyll fluorescence as a tool for stress detection in plants. In: Lichtenthaler, H. K., (ed.), *Applications of Chlorophyll Fluorescence* (pp. 129–142). Kluwer Academic Publishers. Dordrecht, the Netherlands.

Lichtenthaler, H. K., (1996). Vegetation stress: An introduction to the stress concept in plants. *J. Plant Physiol.*, 148, 4–14.

Loreto, F., & Bongi, G., (1989). Combined low temperature-high light effect on gas-exchange properties of jojoba leaves. *Plant Physiol., 91*, 1580–1585. doi: 10.1104/pp.91.4.1580.

Mazza, C. A., Boccalandro, H. E., Giordano, C. V., Battista, D., Sopel, A. L., & Ballare, C. L., (2000). Functional significance and induction by solar radiation of ultraviolet absorbing sunscreens in field grown soybean crops. *Plant Physiology, 122*, 117–125.

Mittler, R., (2006). Abiotic stress, the field environment and stress combination. *Trends Plant Sci., 11*, 15–19. doi: 10.1016/j.tplants.2005. 11.002.

Narendra, S., Venkataramani, S., Shen, G., Wang, J., Pasapula, V., Lin, Y., et al., (2006). The *Arabidopsis* ascorbate peroxidase 3 is a peroxisomal membrane bound antioxidant enzyme and is dispensable for Arabidopsis growth and development. *J. Exp. Bot., 57*, 3033–3042. doi: 10.1093/jxb/erl060.

Ortega-Villasante, C., Rellán-Álvarez, R., Del Campo, F. F., Carpena-Ruiz, R. O., & Hernández, L. E., (2005). Cellular damage induced by cadmium and mercury in *Medicago sativa. J. Exp. Bot., 56*, 2239–2251.doi: 10.1093/jxb/eri223.

Pandey, P., Irulappan, V., Bagavathiannan, M. V., & Kumar, M. S., (2017). Impact of combined abiotic and biotic stresses on plant growth and avenues for crop improvement by exploiting physio-morphological traits. *Frontiers in Plant Science, 8*. Article 537.

Patterson, D. T., (1995). Effects of environmental stress on weed/crop interaction. *Weed Sci., 43*, 483–490.

Pehlivan, E., Özkan, A. M., Dinc, S., & Parlayici, S., (2009). Adsorption of Cu2+ and Pb2+ ion on dolomite powder. *J. Hazard. Mater., 167*, 1044–1049. doi: 10.1016/j.jhazmat.2009.01.096.

Prasad, P. V. V., Pisipati, S. R., Momcilovic, I., & Ristic, Z., (2011). Independent and combined effects of high temperature and drought stress during grain filling on plant yield and chloroplast EF-Tu expression in spring wheat. *J. Agron. Crop Sci., 197*, 430–441. doi: 10.1111/j.1439-037X.2011.00477.x.

Rampino, P., Pataleo, S., Gerardi, C., & Perotta, C., (2006). Drought stress responses in wheat: Physiological and molecular analysis of resistant and sensitive genotypes. *Plant Cell Environ., 29*, 2143–2152. doi: 10.1111/j.1365- 3040.2006.01588.x.

Samarah, N. H., (2005). Effects of drought stress on growth and yield of barley. *Agron. Sustain. Dev., 25*, 145–149. doi: 10.1051/agro:2004064.

Scandalios, J. G., (2005). Oxidative stress: Molecular perception and transduction of signals triggering antioxidant gene defenses. *Braz. J. Med. Biol. Res., 38*, 995–1014.doi: 10.1590/S0100-879X2005000700003.

Selye, H., (1936). A syndrome produced by various nocuous agents. *Nature, 138*, 32-34.

Sharma, A., Shahzad, B., Rehman, A., Bhardwaj, R., Landi, M., & Zheng, B., (2019). Response of phenylpropanoid pathway and the role of polyphenols in plants under abiotic stress. *Molecules, 24*, 2452. doi: 10.3390/ molecules24132452.

Sharma, P., Jha, A. B., Dubey, R. S., & Pessarakli, M., (2012). Reactive oxygen species, oxidative damage, and antioxidative defense mechanism in plants under stressful conditions. *J. Bot., 2012*, 1–26. doi: 10.1155/2012/2 17037.

Sharma, S., chatterjee, S., Kataria, S., Joshi, J., Datta, S., Vairale, M. G., & Veer, V., (2017). *A Review on Responses of Plants to UV-B Radiation Related Stress*. John Wiley & Sons Ltd. Published by John Wiley & Sons Ltd.

Su, L., Dai, Z., Li, S., & Xin, H., (2015). A novel system for evaluating drought– cold tolerance of grapevines using chlorophyll fluorescence. *BMC Plant Biol., 15*, 82. doi: 10.1186/s12870-015-0459-8.

Suzuki, N., Rivero, R. M., Shulaev, V., Blumwald, E., & Mittler, R., (2014). Abiotic and biotic stress combinations. *New Phytol., 203*, 32–43. doi: 10.1111/nph. 12797.

Tangahu, B. V., Sheikh, A. S. R., Basri, H., Idris, M., Anuar, N., & Mukhlisin, M., (2011). A review on heavy metals (As, Pb, and Hg) uptake by plants through phytoremediation. *Int. J. Chem. Eng., 2011*, 31. doi: 10.1155/2011/ 939161.

Teramura, A. H., Ziska, L. H., & Sztein, A. E., (1991). Changes in growth and photosynthetic capacity of rice with increased UV-B radiation. *Physiologia Plantarum, 83*, 373–380.

Traunfeld, J. H., & Clement, D. L., (2001). *Lead in Garden Soils*. Home and Garden, Maryland Cooperative Extension. College Park, MD: University of Maryland.

Tripathi, R. D., Srivastava, S., Mishra, S., Singh, N., Tuli, R., Gupta, D. K., et al., (2007). Arsenic hazards: Strategies for tolerance and remediation by plants. *Trends Biotechnol., 25*, 158–165. doi: 10.1016/j.tibtech.2007.02.003.

Valerio, M., Lovelli, S., Perniola, M., Di Tommaso, T., & Ziska, L., (2013). The role of water availability on weed–crop interactions in processing tomato for southern Italy. *Acta Agric. Scand. Sect. B, 63*, 62–68. doi: 10.1080/09064710.2012. 715184.

Van, Z. P. A., Tobler, M. A., Mouton, E., Hasenstein, K. H., & Mopper, S., (2003). Positive and negative consequences of salinity stress for the growth and reproduction of the clonal plant, *Iris hexagonal. J. Ecol., 91*, 837–846. doi: 10.1046/j.1365-2745.2003.00806.x.

Verma, P. K., Verma, S., Pande, V., Mallick, S., Tripathi, R. D., Dhankher, O. P., et al., (2016). Overexpression of rice glutaredoxin OsGrx_C7 and OsGrx_C2. 1 reduces intracellular arsenic accumulation and increases tolerance in *Arabidopsis thaliana. Front. Plant Sci., 7*, 740.

Verma, S., Verma, P. K., Meher, A. K., Bansiwal, A. K., Tripathi, R. D., & Chakrabarty, D., (2017). A novel fungal arsenic methyltransferase, WaarsM reduces grain arsenic accumulation in the transgenic rice plant. *J. Hazard. Mater.*, *344*, 626–634. doi: 10.1016/j.jhazmat.2017.10.037.

Welfare, K., Yeo, A. R., & Flowers, T. J., (2002). Effects of salinity and ozone, individually and in combination, on the growth and ion contents of two chickpea (*Cicer arietinum* L.) varieties. *Environ. Pollut.*, *120*, 397–403. doi: 10.1016/S0269-7491(02)00109-4.

Yun, K. Y., Park, M. R., Mohanty, B., Herath, V., Xu, F., Mauleon, R., et al., (2010). Transcriptional regulatory network triggered by oxidative signals configures the early response mechanisms of japonica rice to chilling stress. *BMC Plant Biol.*, *10*, 16. doi: 10.1186/1471-2229-10-16.

Zhang, W. H., & Tyerman, S. D., (1999). Inhibition of water channels by $HgCl_2$ in intact wheat root cells. *Plant Physiol.*, *120*, 849–858. doi: 10.1104/pp.120.3.849.

Zhou, Z. S., Huang, S. Q., Guo, K., Mehta, S. K., Zhang, P. C., & Yang, Z. M., (2007). Metabolic adaptations to mercury-induced oxidative stress in roots of *Medicago sativa* L. *J. Inorg. Biochem.*, *101*, 1–9. doi: 10.1016/j.jinorgbio.2006. 05.011.

Ziska, L. H., Tomecek, M. B., & Gealy, D. R., (2010). Evaluation of competitive ability between cultivated and red weedy rice as a function of recent and projected increases in atmospheric CO_2. *Agron. J.*, *102*, 118–123. doi: 10.2134/ agronj2009.0205.

CHAPTER 8

Rhizosphere Microbiome Manipulation

MUHAMMAD SIDDIQUE AFRIDI,[1] ABDUL SALAM,[2] and
FLAVIO HENRIQUE VASCONCELOS DE MEDEIROS[1]

[1]*Department of Plant Pathology, Federal University of Lavras, Lavras, MG, Brazil*

[2]*Zhejiang Key Lab of Crop Germplasm, Department of Agronomy, College of Agriculture and Biotechnology, Zhejiang University, Hangzhou, China*

ABSTRACT

Rhizosphere microbiome's biodiversity and functionality play a key role in biogeochemistry and plant performance. Rhizosphere ecosystem harbors diverse microbial communities which underpin plant growth, fitness via employing direct and indirect mechanisms under adverse conditions. The utmost roles of the rhizosphere microbiome are plant growth-promotion, bio-fertilization, bio-stimulation, and biocontrol of plant disease. With the advent of advanced technologies such as metagenomics and transcriptomics, unraveled the complexities, structures, and core functions of microbial communities and understood the interactions between microbes and host in the rhizosphere. In this chapter, we have discussed rhizosphere microbiome diversity, interaction, and its influence on plant health, fitness, and protection. Of note, we also highlighted the potential strategies of rhizosphere microbiome manipulation that could be harnessed for rhizosphere microbiome engineering to improve plant health and agriculturally sustainability.

Advances in Plant Microbiome Research for Climate Resilient Agriculture: Toward Sustainable Farming.
Ashwani Kumar (Ed.)

8.1 INTRODUCTION

The rhizosphere has been defined as the vibrant zone of soil influenced by the root activity. This influence is caused by the release of plant's organic matter, a phenomenon known as rhizosphere deposition, which consists mainly of plant metabolites (secretions/exudates) and plant debris (Dessaux, Grandclément, & Faure, 2016). It is a narrow but dynamic region of the soil where substantial biotic and abiotic interactions exist between roots and the soil-borne microbiota (Mueller & Sachs, 2015). Rhizosphere soils promoting the growth of microbial communities such as bacteria, microbial viruses, fungi, and archaea are termed as mesotrophic environment while the bare soils are regarded as oligotrophic soil (Dessaux, Grandclément, & Faure, 2016). Plant-associated microorganisms and microbiomes afford vital 'ecosystem services' by mitigating biotic and abiotic stresses and enhancing plant growth (Arif, Batool, & Schenk, 2020). Rhizosphere consists of three distinct layers, i.e., (i) endo-rhizosphere; the area where the activity of the rhizosphere is most intense (the outer layer on the surface of the plant roots); (ii) rhizoplane; the actual root-soil crossing point (the central zone, the root epidermis and the internal layer in connection with the mucilage); and (iii) ecto-rhizosphere; the outermost layer of the rhizosphere (Ryan et al., 2009).

Contrarily, it is projected that the global population will reach nearly 9.6 billion by 2050, thus the agricultural production needs to be increased by 70 to 100% to feed the growing population (Rodrigues et al., 2017; Zhao et al., 2020; Mueller et al., 2012). On the other side, land degradation and rapid pace of urbanization put pressure on arable land (Munns et al., 2012). The rapidly increasing population adds significant strains on arable land and agriculture sustainability. Global climatic conditions further worsened the problem, imposing extra encumbrance on crop production. The situation became alarming in recent past years (Raza et al., n.d.). For instance, new technologies needed to be arrayed in agriculture to achieve crop sustainability and higher yield production. The green revolution has significantly increased agricultural output via genetically modified crops, chemical fertilizers and pesticides, but this success has been hampered by many unsustainable practices, such as degraded soil health (Arif, Batool, & Schenk, 2020). Although genetically manipulated plants amongst them remains a sustainable option but the research and regulatory costs required to drive new varieties are costly and can take years to expand it to the consumers and get regulatory approval (Syed et al., 2018). In contrast, manipulating rhizosphere microbiome is a low-cost, an alternative emerging tool and a sustainable practice to increase crop yields and resilience. The key goal of rhizosphere microbiome manipulation is to reshape its ecology and physiology in favor of plants and protect its host from both biotic and abiotic stresses and increase plant yield. So far, the rhizosphere microbiome has only been studied and focused on the characterization of bacterial and fungal communities (Quiza, St-Arnaud, & Yergeau, 2015; Kumar & Dubey, 2020), while the rest of the communities, such as nematodes (fungivorous nematodes), protists, and earthworms are entirely ignored. Collectively, these diverse microbe-animal communities have beneficial impacts on their hosts, such as enhancing plant growth and protecting it under biotic and abiotic stresses (Rosenberg & Zilber-Rosenberg, 2013). Manipulating the implausible host plant-associated microbial diversity is a steppingstone to elucidating the structure and gene function of microbiota. Here, the techniques mentioned below have been used to manipulate/engineer beneficial microbiome functions and services to ensure the productivity of plants across multiple generations.

8.2 RHIZOSPHERE MICROBIOME DIVERSITY

Plant associated microbial communities are diverse and abundant in numbers, play crucial roles in plant health and protection, thus it is also called the second genome of the plant (Berendsen, Pieterse, & Bakker, 2012).

The soil microbial community is by far the largest known pool of biodiversity in the world, and the rhizosphere is the narrow region of the soil affected by root exudates that can contain as many as 10^{11} microbial cells per gram of root (Egamberdieva et al., 2007) and more than 30,000 prokaryotes species (Figure 8.1) (Curtis et al., 2002). The mass genome of this microbial community is much larger than that of the plant and is also called the second genome of the plant. However, microbial diversity is inclined by the physical and chemical properties of the rhizosphere, some of which are determined by the host plant's genetics (Peiffer et al., 2013).

Rhizosphere microbiome is one of the complex and diverse hot spots that harbors both beneficial and harmful microbes which closely associate to plant roots and significantly influence host health. This soil borne plant pathogens cause devastating disease and potentially damage host plant and it is estimated that 20–30% of agriculture productivity is only lost due to soil-borne pathogens globally (Peiffer et al., 2013; Savary et al., 2019). Yet, it's still a barrier, and the scientific communities rely on cultivation-dependent techniques to identify that particular group of microorganisms responsible for disease suppressiveness (Berendsen et al., 2012).

Rhizosphere microbiomes possess diverse groups of microbial communities. These numerous groups play substantial role in plant health, nutrients recycling and pathogen suppression (Figure 8.1). Bunches of evidence are reported that rhizosphere microbiome has the potential to promote plant growth and can be engineered for high production (Vannier, Agler, & Hacquard, 2019; Chaparro, Badri, & Vivanco, 2013, 2014).

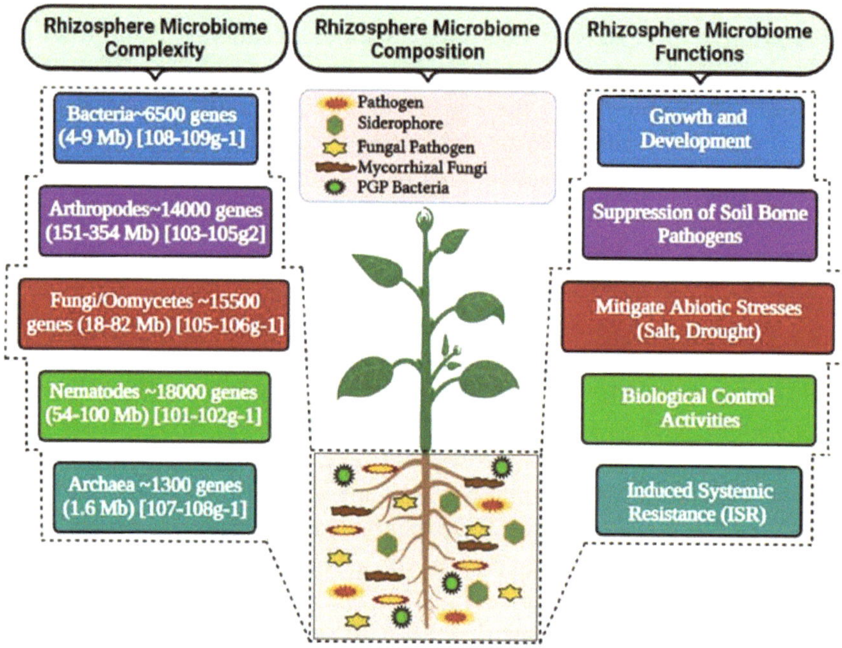

FIGURE 8.1 Rhizosphere overview of rhizosphere microbiome diversity, composition, and functions.

8.3 RHIZOSPHERE MICROBIOME INTERACTION

The thin zone surrounded by plant roots nurture diverse groups of microbes and invertebrates perform crucial role in nutrients mobilization and directly influenced by the plant roots referred as rhizosphere (Figure 8.1) (York et al., 2016). These microbial communities establish association with host plant and constantly interacting with one another via different association in order to guarantee their survival. These microorganisms and animal exhibit different association between one another and host plant such as commensalism, parasitism, amensalism, saprophytism, and symbiosis. These mutual associations among microbes and with plants influence plant health and fitness (Mendes et al., 2011; Berendsen et al., 2012).

Root exudates majorly serve as signals to recruit microbes to the rhizosphere irrespective of microbe species, therefore, Rhizosphere microbiome harbors both pathogenic and beneficial microbes (Compant, Samad, Faist, & Sessitsch, 2019; Rodriguez et al., 2019) although, it has been reported that plant recruits beneficial microbes through root exudates that assist plants under unfavorable circumstances (Berendsen et al., 2012), yet these microorganisms differ in traits and phylogeny, they co-exist and operate together despite their divergences. The complexity of the soil and its richness in nutrients makes it a favorable environment for various forms of life to exist microorganism interactions, co-existence, and "divergence-overlook" in the rhizosphere are controlled by signals released from plants through the roots into the rhizosphere. These are low and high-weight molecular signaling compounds called exudates (Berendsen et al., 2012).

Rhizosphere microbiome harbors many complex interacting organisms that drive functional activities that have a positive impact on plant health. Bacteria, fungi, viruses, and archaea, among others, are present in the rhizosphere. These organisms interact together and their interaction has an effect on the rhizobiome. Quorum signaling (QS) is a well-established communication and coordination tool that microbes used for microbe-microbe interactions. For instance, gram-negative bacteria, produce N-acyl-l-homoserine lactones (AHLs), release into the rhizosphere and the plant triggered to produce metabolites in the response of AHLs (Hassani et al., 2019). Moreover, microbes also employ chemotaxis phenomenon in response to plant signals (i.e., sugars or organic acids), to colonize host root to establish plant microbes' interactions (O'Neal, Vo, & Alexandre, 2020). QS facilitate finger millet endophyte M6 (Enterobacter sp.) in plant colonization and biofilm formation on rot hairs to grab and prevent the endophytic pathogen into plant tissue (Liu et al., 2020).

8.4 THE POTENTIAL ROLE OF RHIZOSPHERE MICROBIOME ON PLANT HEALTH

Plant associated microbes execute numerous processes needed for plant health and fitness, thus, collectively microbial functions play key role in sustainable crop production.

Beneficial microbes are the volunteer engines of soil fertility, and plant performance basically depends on array of functions exhibited by soil and root associated microbes. For example, beneficial bacteria and fungi provide unavailable soil nutrients and mobilize them to plants by secreting organic and inorganic acids, N_2 fixation, phytohormones siderophores and enzymes production (Rahimi et al., 2020; Ali et al., 2022). They help plants to cope the abiotic stresses by regulating their phytohormones and guarding them against diverse groups of enemies such as insect pests, parasites, and pathogens (Figure 8.1) (Berendsen et al., 2012; Fierer, 2017). Moreover, an intricate existed interaction between host and its microbiome triggers the host immunity which consequently induce tolerance in host heterogeneous environments (Teixeira et al., 2019).

Plant microbiome is dynamic; its compositions and active functions are constantly change with regards to host developmental stages and environmental stimuli (Timm et al., 2018; Santos-Medellín et al., 2017). These external changes significantly influence microbial activates and assembly which consequently impact plant health and performances (Durán et al., 2018; Chen et al., 2018). Under these unfavorable conditions, plant adapt strategy what was coined "cry for help" (Bakker et al., 2018), when plant encounters multiple a(biotic) stresses or pest induced stresses, recruit beneficial microbiomes from the surrounding by secreting numerous chemical stimuli to increase their potential to combat stresses (Bakker et al., 2018; Carrión et al., 2019) but this is also undeniable reality, that innate immunity of plant is not enough alone to guard host from enemies in rhizosphere or phyllosphere. These microbial communities inhabiting in rhizosphere, phyllosphere potentially guard host from biotic invaders and assist under environmental stresses (Durán et al., 2018; Carrión et al., 2019).

Plant microbiome functions also link the host genomics and reduce the biotic stresses through triggering defense related genes which support plant defense-related metabolic pathways. This statement got strengthened when Chialva et al. (2018) reported in his studies that plant defense system primed through microbe-associated molecular patterns (MAMPs) of various microbes belonging to microbiome (Hubbard et al., 2019).

Microbiome potential capacity regarding plant health and protection may not apply in all soils as their composition, chemistry, and biology are significantly different from one another at different geographical locations, this is how the microbiome functions and potential are still unclear to what extent microbiomes prim and induce tolerance in host plant against both biotic and abiotic stresses (Chialva et al., 2018).

8.5 RHIZOSPHERE MICROBIOME MANIPULATION

The rapid evaluation of microbiome and its mesmerized potential for plant growth and development caught the interest of scientific communities in previous couple of decades. This evaluation brought up new hopes for sustainable agriculture production.

Rhizosphere microbiome composed of diverse microbial communities which underpin plant in acquiring soil nutrients and suppressing soil-borne pathogens. This interaction between plant and surrounded microorganisms is established through roots exudates. These roots exudates intrigues nearby soil microbes. The variation exists in microbial communities, however, between plant species and even between genotypes (York et al., 2016).

It is unquestioned that microbiome is not static and recognizing the drivers which influence rhizosphere microbiome are challenging. However, manipulation could achieve the applicable goals in a sustainable agriculture system.

Among them, agroecosystem is highly intricate and diverse target to be manipulated as it encompasses massive mass of microorganisms and entangled webs of interactions. Plant, microbes, insects are the major compounds of it (Berendsen et al., 2012; Busby et al., 2017) (such as fungi, archaea, bacteria, and protists; Figure 8.1). Researchers are now faced with the unprecedented task of exploiting the functions of these microbial communities (microbiomes), as new pest/pathogen strains, climate change and limited availability of chemical fertilizers are progressively intimidating agricultural production stability. Achieving this goal of worldwide doubling the agricultural production by 2050 needs immediate solution (Tilman, Balzer, Hill, & Befort, 2011; Callaway, 2016; Dangl, Horvath, & Staskawicz, 2013).

Manipulation of rhizosphere microbiome employing classical and biotechnological strategies improve plant growth and induce tolerance to biotic and abiotic stresses could be the untapped potential of rhizosphere microbiome tackling sustainable production in agriculture.

8.5.1 SOIL AMENDMENTS

Soil amendments include the addition of organic and inorganic substances to soils or agricultural practices to increase agricultural production, influence, and engineer microbiome plant interactions (Arif, Batool, & Schenk, 2020). Organic amendments include the addition of various additives to the soil as an energy source for plant-beneficial heterotrophic bacteria, fungi, and invertebrates (Sankar et al., 2017). They have many benefits proliferating with substantial alteration in depleted soil fertility, plant and microbial biomass, root attributes, elemental composition of the tissue, substrate use, decomposition, and reducing farming waste (Sankar et al., 2017; Ganesh, Sridhar, & Vishali, 2022). Organic modification triggers a series of events between different ecosystem trophic levels by modifying the biophysical properties of the soil, heterogeneous distribution of microbes, and plant characteristics. Implementing organic modifications affects the structure, composition, and distribution of microbial community (Cesarano et al., 2017).

Inorganic modifications including perlite, vermiculite, lime, and sand that are used to decrease the toxic effects of soil salinity and acidity on crops (Singh et al., 2019). They can also lead to increase plant biomass, which results in increasing root exudates, and a larger more active microbiome. The combination of both inorganic and organic amendments resulted a 30% increase in crop yield compared to an 8% increase with only organic fertilizers (Ai et al., 2015).

8.5.2 COVER CROP

A vital portion of the earth's biodiversity is underneath ground (Bardgett & Van, 2014), a substantial part of soil microbial communities resides in rhizosphere or rhizomicrobiome. It is considered one of the most intricate and functionally active ecosystems on earth, which boosts plant health and alleviates biotic and abiotic stresses.

Rhizosphere microbiome harbors diverse groups of microbes such as bacteria, oomycetes, fungi, archaea, and viruses. These different groups interact with one another, and established association with host through direct and indirect interaction (Bandyopadhyay et al., 2017; Lakshmanan, Selvaraj, & H. B. P., 2014).

This diversification of microbiome plays key role in soil health, biochemical process and structure composition (E. B., 2007). To meet the food demands for growing human population, the extensive use of agrochemicals and some conventional agricultural practices such as monoculture (Raaijmakers et al., 2009) plant residues lifting (Baird, Watson, & Scruggs, 2003) are practiced in agriculture system which significantly increased and fulfilled the food requirements. But on the other hand, these practices potentially transform the compositions of microbial communities and imbalanced the multifunctionality of rhizosphere microbiome.

In last two decades, massive research has been reported which indicates that plant diversity link to natural ecosystem function, diverse microbial communities that lead the promotions of ecosystem functioning and soil multifunctionality (Zuo & Zhang, 2011; Santoyo, Orozco-Mosqueda, Govindappa, 2012; Delgado-Baquerizo et al., 2016; Wagg et al., 2014). Plant associated microbes perform substantial roles in maintaining ecosystems functions, improve plant growth, pathogen control and tolerance under drastic environmental constraints (Van et al., 2008; Haney et al., 2015; Berg, Rybakova, & M. G., 2016; Wagg et al., 2014). To attain the desired goals without being compromised om rhizosphere microbiome multifunctions and diversity, rhizosphere microbiome could be harnessed via cover cropping agriculture practice.

Cover cropping system improve soil fertility, upsurge microbial diversity and promote plant growth without altering other facets of agricultural practices radically (Schipanski et al., 2014). For instance, recently Garland et al. (2021) assessed cover cropping and rotational system in his studies and concluded that cover crop is more important than crop rotation as it increased soil microbial diversity, multifunctions, and crop production in European cropping systems. It has been linked with enhanced soil fertility, soil microbial diversity, increased soil tilt and aggregate stability, better soil water management, and reduced erosion (Larkin et al., 2021). Cover crop rotation positively affect the diversity and richness of fungal community (Li & Q. L., 2019). To manipulate the rhizosphere microbiome for better plant growth and productivity, cover crop strategy is an ecofriendly, sustainable, and accessible approach in agriculture system, hence, this practice is not effective for all agronomic crops and a large body of research needed to investigate its effectiveness across the different agronomic crops.

8.5.3 INTRODUCTION OF PLANT BENEFICIAL MICROBES

The rhizosphere is a playground for diverse microbial communities where they interact with one another and with host which consequently influence plant health, by alleviating a(biotic) stress conditions. Plant associated beneficial microbes underpin host by providing soil nutrients, suppressing the phytopathogens and induce stress resilience under harsh environments have been massively documented (Afridi et al., 2019; Mehmood et al., 2021; ALKahtani et al., 2020; Alok et al., 2020). Harnessing beneficial microbes' functions for rhizosphere microbiome manipulation marked as a key strategy in agriculture system (Baird, Watson, & Scruggs, 2003).

Most of the rhizosphere microbiome engineering strategies underlying microbes' culture to escalate cultivability potential of rhizospheric microbes which exhibit specific functional capability, but how these microorganisms would react towards exogenous environmental conditions are unknown (Dubey et al., 2019; Kumar, Vyas, Malla, & Dubey, 2019).

Beneficial microbes as bioinoculants have been frequently applied with the aim to increase nutrients arability, improve plant growth and induce stress tolerance under extreme environmental conditions (Alok et al., 2020). These different microbial groups also serve as competent biocontrol agents which safeguard host against enemies by producing various antibiotic compounds such as phenazine, DAPG, HCN, oligomycin, bacteriocines (Nisin) along with the production of antifungal compounds including phoroglucinols, and pyoluteorin (Dubey et al., 2019; Hashem et al., 2017; Khan et al., 2019).

The introduction of exogenous strains to manipulate microbiomes are usually unable to compete with the native microbiota and eliminated in soils have been extensively reported in various studies, while it has been also documented that inoculation of PGPB significantly robust plan growth and productivity (Afridi et al., 2021; Ali et al., 2022; Finkel et al., 2017). This contradiction questioned the importance of PGPB application in rhizosphere microbiome manipulation. But recently Chen et al. (2022) resolved this puzzle and reported in his study that inoculated PGPB induced DNA methylation in roots which consequently improve plant growth, its related promotion process and these amendments remained purposeful after microbiome inoculum elimination.

This research displays novel worthwhile trait OF PGPB being inoculated and open new window for rhizosphere microbiome manipulation beyond the perception of upholding inoculum persistence in soils.

8.6 HOST-MEDIATED MICROBIOME ENGINEERING

Host-mediated microbiome engineering (HMME) is promising biotechnological tool/strategy using host phenotype for indirectly microbiota selection via cyclic differentiation and propagation (Panke-Buisse et al., 2015). This strategy visualize phenotypically changes in host plant and allow to recruit and select specific microbiome from the growing zone after various generation and propagation (Qiu et al., 2019). The host phenotype is modified with the objective to manipulate the rhizosphere microbiome to select the beneficial microbes which influence plant health, fitness, and underpin under unfavorable conditions (Mueller & Sachs, 2015; French et al., 2021). It has been well documented in various studies that plant secrets low molecular weight and high molecular weight compounds known are rot exudates performs various functions such as signaling broadcasting, microbial activities and mediation of plant microbe interaction in rhizosphere (Akiyama, Matsuzaki, & Hayashi, 2005; Chaparro et al., 2013; Baetz & Martinoia, 2014). These exudates reshape and assembles microbial communities and architecture a healthy and resilient soil microbiome to protect itself from disease and increase productivity (Vannier, Agler, & Hacquard, 2019; Berg, & Smalla, 2009). This novel approach has been shown promising results and counteracted the abiotic drought stress effect when applied (Jochum et al., 2019).

Plant produces a diverse group of exudates and their composition varies by plant species, and even cultivars within a species (Kowalchuk et al., 2002; Micallef et al., 2009) which consequently each plant spp. or genotype possess different microbiome (Kuklinsky-Sobral et al., 2004; Salles, Van, & Van, 2004). These variations of root exudate among plant species and genotypes recommends the promising potential of root exudates that could be manipulate an agriculture genotype for desired selective effect by plant breeding to engineer the objective rhizosphere microbiome for plant health and productivity.

8.7 MICROBIOME BREEDING

Microbiome engineering seeks to alter the composition or activity of complex microbial populations. Numerous of studies since last three decades have clearly demonstrated that plant genotype and soil type are the two main drivers shaping the rhizosphere microbiome (Berendsen et al., 2012; Berg & Smalla, 2009). Ample of approaches have been carried out to manipulate the microbiome to boost plant growth, yield, and resilience. Microbiome breeding is one of the promising emerging biotechnological strategies that could be harnessed for rhizosphere microbiome manipulation (Arif, Batool, & Schenk, 2020). This strategy totally based on the host plant that exhibits specific traits that supports and establish association with the beneficial microbiota through generations. The host indirectly select the microbiome by recruiting the subset of microbial population and allowing them to interact and then transmit vertically to their offspring (Mueller & Sachs, 2015). Recently, Dessaux et al. (2016) reviewed in his articles "Engineering the Rhizosphere" that Microbial Partners who occupied the root system and strongly associate with host could be engineered and render them more effectively. This concept has been supported in one of the studies in which a heterologous gene that encodes siderophore receptor introduced into a *Pseudomonas fluorescens* strain to strengthens his competitiveness in soil (Raaijmakers et al., 1995). Recently, Zhang et al. (2012) used his efforts and introduced chitinase gene from *Bacillus subtilis* into the *Burkholderia vietnamiensis* PGPR strain P418 successfully and significantly suppressed the cryptogamic diseases such as wheat sheath blight, cotton Fusarium wilt, and tomato gray mold.

Microbiome breeding strategy has diverse potential application such as rhizosphere microbiome breeding for plant growth promotion or to show resilience for biotic and abiotic stresses. However, to ensure its competitiveness, the microbial ecology, functions, longevity, and their response between and with plants under changing climatic conditions should be study under diverse field environments.

8.8 CONCLUSION

The rhizosphere microbiome harbors diverse microbial communities, declared as a powerful tool for reducing the usage of agrochemicals, combating plant pathogens, and upraising crop yield in a sustainable agriculture system. The key function of active microbiome to underpin plant growth and development and induce stress tolerance under adverse environmental conditions. The fundamental interactions and underlying mechanisms in rhizosphere microbiome have been revealed through experimental biology, multi-omics, and computational biology approaches. Recently, rhizosphere microbiome engineering for plant growth and productivity insinuated in multiple studies. Diverse groups of beneficial microbes have been applied as bio-inoculants in field conditions and achieved the desired goals. Still, these groups encountering the barriers of a(biotic) constraints and failed to give maximum output in field conditions. Therefore, further studies are needed to discover the influence of a(biotic) factors on these inoculated beneficial microbes in rhizosphere microbiome environment. Advanced omics tools and molecular based approaches will undoubtedly ease to attain the goal, but at the same time, we are unable to overlooked and must appreciate the significant role of conventional strategies in a healthy rhizosphere microbiome manipulation. To optimize and enhance the positive functions of rhizosphere microbiome for plat growth and productivity, integrated of multitude studies are required to harnesses both conventional and advanced approach for rhizosphere microbiome manipulation. Advanced knowledge about the dynamics of rhizosphere microbiome and the effect of both biotic and abiotic factors on its compositions and functions will pave the way for the rhizosphere microbiome manipulation for sustainable agriculture production under harsh conditions.

KEYWORDS

- **beneficial microbes**
- **microbiome manipulation**
- **plant beneficial microbes**
- **plant health**
- **rhizosphere**
- **soil amendments**

REFERENCES

Afridi, M. S., Amna, Sumaira, Mahmood, T., Salam, A., Mukhtar, T., et al., (2019). Induction of tolerance to salinity in wheat genotypes by plant growth promoting endophytes: Involvement of ACC deaminase and antioxidant enzymes. *Plant Physiol Biochem., 139*, 569–577. https://doi.org/10.1016/j.plaphy.2019.03.041.

Afridi, M. S., Van, H. J D., Bundschuh, J., Sumaira, Khan, M. N., Salam, A., et al., (2021). Biotechnological approaches in agriculture and environmental management – bacterium *Kocuria rhizophila* 14ASP as heavy metal and salt- tolerant plant growth- promoting strain. *Biologia (Bratisl);76*, 3091–3105. https://doi.org/10.1007/S11756-021-00826-6/FIGURES/5.

Ai, C., Liang, G., Sun, J., Wang, X., He, P., Zhou, W., et al., (2015). Reduced dependence of rhizosphere microbiome on plant-derived carbon in 32-year long-term inorganic and organic fertilized soils. *Soil Biol Biochem., 80*, 70–78. https://doi.org/10.1016/J.SOILBIO.2014.09.028.

Akiyama, K., Matsuzaki, K., & Hayashi, H., (2005). Plant sesquiterpenes induce hyphal branching in arbuscular mycorrhizal fungi. *Nature, 435*, 824–827. https://doi.org/10.1038/nature03608.

Ali, B., Wang, X., Saleem, M. H., Azeem, M. A., Afridi, M. S., Nadeem, M., et al., (2022). *Bacillus* mycoides PM35 reinforces photosynthetic efficiency, antioxidant defense, expression of stress-responsive genes, and ameliorates the effects of salinity stress in maize. *Life., 12*, 219. https://doi.org/10.3390/life12020219.

Ali, B., Wang, X., Saleem, M. H., Sumaira, Hafeez, A., Afridi, M. S., et al., (2022). PGPR-mediated salt tolerance in maize by modulating plant physiology, antioxidant defense, compatible solutes accumulation and bio-surfactant producing genes. *Plants, 11*, 345. https://doi.org/10.3390/PLANTS11030345.

ALKahtani, M. D. F., Fouda, A., Attia, K. A., Al-Otaibi, F., Eid, A. M., El-Din, E. E., et al., (2020). Isolation and characterization of plant growth promoting endophytic bacteria from desert plants and their application as bioinoculants for sustainable agriculture. *Agron., 10*, 1325. https://doi.org/10.3390/AGRONOMY10091325.

Alok, D., Annapragada, H., Singh, S., Murugesan, S., & Singh, N. P., (2020). Symbiotic nitrogen fixation and endophytic bacterial community structure in Bt-transgenic chickpea (*Cicer arietinum* L). *Sci. Rep., 10*, 5453. https://doi.org/10.1038/s41598-020-62199-1.

Arif, I., Batool, M., & Schenk, P. M., (2020). Plant microbiome engineering: Expected benefits for improved crop growth and resilience. *Trends Biotechnol., 38*, 1385–1396. https://doi.org/10.1016/j.tibtech.2020.04.015.

Baetz, U., & Martinoia, E., (2014). Root exudates: The hidden part of plant defense. Trends *Plant Sci., 19*, 90–98. https://doi.org/10.1016/j.tplants.2013.11.006.

Baird, R. E., Watson, C. E., & Scruggs, M., (2003). Relative longevity of *Macrophomina phaseolina* and associated mycobiota on residual soybean roots in soil. *Plant Dis., 87*, 563–566. https://doi.org/10.1094/PDIS.2003.87.5.563.

Bakker, P. A. H. M., Pieterse, C. M. J., De Jonge, R., & Berendsen, R. L., (2018). The soil-borne legacy. *Cell, 172*, 1178–1180. https://doi.org/10.1016/J.CELL.2018.02.024.

Bandyopadhyay, P., Bhuyan, S. K., Yadava, P. K., Varma, A., & Tuteja, N., (2017). Emergence of plant and rhizospheric microbiota as stable interactomes. *Protoplasma, 254*, 617–626. https://doi.org/10.1007/S00709-016-1003-X.

Bardgett, R. D., & Van, D. P. W. H., (2014). Belowground biodiversity and ecosystem functioning. *Nature, 515*, 505–511. https://doi.org/10.1038/nature13855.

Barrios, E. (2007). Soil biota, ecosystem services and land productivity. *Ecological Economics, 64*(2), 269–285. https://doi.org/10.1016/j.ecolecon.2007.03.004.

Berendsen, R. L., Pieterse, C. M. J., & Bakker, P. A. H. M., (2012). The rhizosphere microbiome and plant health. *Trends Plant Sci., 17*, 478–486. https://doi.org/10.1016/j.tplants.2012.04.001.

Berg, G., & Smalla, K., (2009). Plant species and soil type cooperatively shape the structure and function of microbial communities in the rhizosphere. *FEMS Microbiol. Ecol., 68*, 1–13.

Berg, G., Rybakova, D., Grube, M., & Köberl, M. (2016). The plant microbiome explored: implications for experimental botany. *Journal of Experimental Botany, 67*(4), 995–1002.

Busby, P. E., Soman, C., Wagner, M. R., Friesen, M. L., Kremer, J., Bennett, A., et al., (2017). Research priorities for harnessing plant microbiomes in sustainable agriculture. *PLOS Biol., 15*, e2001793. https://doi.org/10.1371/journal.pbio.2001793.

Callaway, E., (2016). Devastating wheat fungus appears in Asia for first time. *Nature, 532*, 421–422. https://doi.org/10.1038/532421a.

Carrión, V. J., Perez-Jaramillo, J., Cordovez, V., Tracanna, V., De Hollander, M., Ruiz-Buck, D., et al., (2019). Pathogen-induced activation of disease-suppressive functions in the endophytic root microbiome. *Science (80-), 366*, 606–612. https://doi.org/10.1126/science.aaw9285.

Cesarano, G., De Filippis, F., La Storia, A., Scala, F., & Bonanomi, G., (2017). Organic amendment type and application frequency affect crop yields, soil fertility and microbiome composition. *Appl. Soil Ecol., 120*, 254–264. https://doi.org/10.1016/j.apsoil.2017.08.017.

Chaparro, J. M., Badri, D. V., & Vivanco, J. M. (2014). Rhizosphere microbiome assemblage is affected by plant development. *The ISME Journal, 8*(4), 790–803. https://doi.org/10.1038/ismej.2013.196.

Chaparro, J. M., Badri, D. V., Bakker, M. G., Sugiyama, A., Manter, D. K., & Vivanco, J. M., (2013). Root exudation of phytochemicals in *Arabidopsis* follows specific patterns that are developmentally programmed and correlate with soil microbial functions. *PLoS One, 8*, e55731. https://doi.org/10.1371/journal.pone.0055731.

Chen, C., Wang, M., Zhu, J., Tang, Y., Zhang, H., Zhao, Q., et al., (2022). Long-term effect of epigenetic modification in plant–microbe interactions: Modification of DNA methylation induced by plant growth-promoting bacteria mediates promotion process. *Microbiome, 10*, 36. https://doi.org/10.1186/s40168-022-01236-9.

Chen, Y., Wang, J., Yang, N., Wen, Z., Sun, X., Chai, Y., et al., (2018). Wheat microbiome bacteria can reduce virulence of a plant pathogenic fungus by altering histone acetylation. *Nat. Commun., 9*, 3429. https://doi.org/10.1038/s41467-018-05683-7.

Chialva, M., Salvioli, D. F. A., Daghino, S., Ghignone, S., Bagnaresi, P., Chiapello, M., et al., (2018). Native soils with their microbiotas elicit a state of alert in tomato plants. *New Phytol., 220,* 1296–1308. https://doi.org/10.1111/nph.15014.

Chialva, M., Salvioli, D. F. A., Daghino, S., Ghignone, S., Bagnaresi, P., Chiapello, M., et al., (2018). Native soils with their microbiotas elicit a state of alert in tomato plants. *New Phytol., 220*, 1296–1308. https://doi.org/10.1111/NPH.15014.

Compant, S., Samad, A., Faist, H., & Sessitsch, A., (2019). A review on the plant microbiome: Ecology, functions, and emerging trends in microbial application. *J. Adv. Res., 19*, 29–37. https://doi.org/10.1016/j.jare.2019.03.004.

Curtis, T. P., Sloan, W. T., & Scannell, J. W., (2002). Estimating prokaryotic diversity and its limits. *Proc. Natl. Acad. Sci., 99*, 10494–10499. https://doi.org/10.1073/pnas.142680199.

Dangl, J. L., Horvath, D. M., & Staskawicz, B. J., (2013). Pivoting the plant immune system from dissection to deployment. *Science (80), 341*, 746–751. https://doi.org/10.1126/science.1236011.

Delgado-Baquerizo, M., Maestre, F. T., Reich, P. B., Jeffries, T. C., Gaitan, J. J., Encinar, D., et al., (2016). Microbial diversity drives multifunctionality in terrestrial ecosystems. *Nat. Commun., 7*, 10541. https://doi.org/10.1038/ncomms10541.

Dessaux, Y., Grandclément, C., & Faure, D., (2016). Engineering the rhizosphere. *Trends Plant Sci., 21*, 266–278. https://doi.org/10.1016/j.tplants.2016.01.002.

Dubey, A., Kumar, A., Abd_Allah, E. F., Hashem, A., & Khan, M. L., (2019). Growing more with less: Breeding and developing drought resilient soybean to improve food security. *Ecol. Indic., 105,* 425–437. https://doi.org/10.1016/J.ECOLIND.2018.03.003.

Durán, P., Thiergart, T., Garrido-Oter, R., Agler, M., Kemen, E., Schulze-Lefert, P., et al., (2018). Microbial interkingdom interactions in roots promote *Arabidopsis* survival. *Cell, 175*, 973-983.e14. https://doi.org/10.1016/j.cell.2018.10.020.

Egamberdieva, D., Kamilova, F., Validov, S., Gafurova, L., Kucharova, Z., & Lugtenberg, B., (2007). High incidence of plant growth-stimulating bacteria associated with the rhizosphere of wheat grown on salinated soil in Uzbekistan. *Environ. Microbiol., 10*, 070816220133002-??? https://doi.org/10.1111/j.1462-2920.2007.01424.x.

Fierer, N., (2017). Embracing the unknown: Disentangling the complexities of the soil microbiome. *Nat. Rev. Microbiol., 15*, 579–590. https://doi.org/10.1038/nrmicro.2017.87.

Finkel, O. M., Castrillo, G., Herrera, P. S., Salas, G. I., & Dangl, J. L., (2017). Understanding and exploiting plant beneficial microbes. *Curr. Opin. Plant Biol., 38*, 155–163. https://doi.org/10.1016/j.pbi.2017.04.018.

French, E., Kaplan, I., Iyer-Pascuzzi, A., Nakatsu, C. H., & Enders, L., (2021). Emerging strategies for precision microbiome management in diverse agroecosystems. *Nat. Plants, 7*, 256–267. https://doi.org/10.1038/s41477-020-00830-9.

Ganesh, K. S., Sridhar, A., & Vishali, S., (2022). Utilization of fruit and vegetable waste to produce value-added products: Conventional utilization and emerging opportunities: A review. *Chemosphere, 287*, 132221. https://doi.org/10.1016/j.chemosphere.2021.132221.

Garland, G., Edlinger, A., Banerjee, S., Degrune, F., García-Palacios, P., Pescador, D. S., et al., (2021). Crop cover is more important than rotational diversity for soil multifunctionality and cereal yields in European cropping systems. *Nat. Food, 2*, 28–37. https://doi.org/10.1038/s43016-020-00210-8.

Haney, C. H., Samuel, B. S., Bush, J., & Ausubel, F. M., (2015). Associations with rhizosphere bacteria can confer an adaptive advantage to plants. *Nat Plants, 1*, 15051. https://doi.org/10.1038/nplants.2015.51.

Hashem, A., Abd-Allah, E. F., Alqarawi, A. A., Radhakrishnan, R., & Kumar, A., (2017). Plant defense approach of *Bacillus subtilis* (BERA 71) against *Macrophomina phaseolina* (Tassi) Goid in mung bean. *J. Plant Interact., 12*, 390–401. https://doi.org/10.1080/17429145.2017.1373871.

Hassani, M. A., Özkurt, E., Seybold, H., Dagan, T., & Stukenbrock, E. H., (2019). Interactions and coadaptation in plant metaorganisms. *Annu. Rev. Phytopathol., 57*, 483–503. https://doi.org/10.1146/annurev-phyto-082718-100008.

Hubbard, C. J., Li, B., McMinn, R., Brock, M. T., Maignien, L., Ewers, B. E., et al., (2019). The effect of rhizosphere microbes outweighs host plant genetics in reducing insect herbivory. *Mol. Ecol., 28*, 1801–1811. https://doi.org/10.1111/mec.14989.

Jochum, M. D., McWilliams, K. L., Pierson, E. A., & Jo, Y. K., (2019). Host-mediated microbiome engineering (HMME) of drought tolerance in the wheat rhizosphere. *PLoS One*, 14. https://doi.org/10.1371/JOURNAL.PONE.0225933.

Khan, N., Bano, A., Rahman, M. A., Guo, J., Kang, Z., & Babar, M. A., (2019). Comparative physiological and metabolic analysis reveals a complex mechanism involved in drought tolerance in chickpea (*Cicer arietinum* L.) induced by PGPR and PGRs. *Sci Rep., 9*, 2097. https://doi.org/10.1038/s41598-019-38702-8.

Kowalchuk, G. A., Buma, D. S., De Boer, W., Klinkhamer, P. G. L., & Van, V. J. A., (2002). Effects of above-ground plant species composition and diversity on the diversity of soil-borne microorganisms. Antonie van Leeuwenhoek. *Int. J. Gen. Mol. Microbiol., 81*, 509–520. https://doi.org/10.1023/A:1020565523615.

Kuklinsky-Sobral, J., Araújo, W. L., Mendes, R., Geraldi, I. O., Pizzirani-Kleiner, A. A., & Azevedo, J. L., (2004). *Isolation and Characterization of Soybean-Associated Bacteria and Their Potential for Plant Growth Promotion* (Vol. 6, p. 34336706). Wiley Online Libr. https://doi.org/10.1111/j.1462-2920.2004.00658.x.

Kumar, A., & Dubey, A., (2020). Rhizosphere microbiome: Engineering bacterial competitiveness for enhancing crop production. *J. Adv. Res., 24*, 337–352. https://doi.org/10.1016/j.jare.2020.04.014.

Kumar, A., Vyas, P., Malla, M. A., & Dubey, A., (2019). Taxonomic and functional annotation of termite degraded *Butea monosperma* (Lam.) Kuntze (flame of the forest). *Open Microbiol J., 13*, 154–163. https://doi.org/10.2174/1874285801913010154.

Lakshmanan, V., Selvaraj, G., & Bais, H. P. (2014). Functional soil microbiome: belowground solutions to an aboveground problem. *Plant Physiology, 166*(2), 689–700. https://doi.org/10.1104/pp.114.245811.

Larkin, R. P., Griffin, T. S., Honeycutt, C. W., Olanya, O. M., & He, Z., (2021). Potato cropping system management strategy impacts soil physical, chemical, and biological properties over time. *Soil Tillage Res., 213*, 105148. https://doi.org/10.1016/J.STILL.2021.105148.

Li, W. H., & Liu, Q. Z. (2019). Changes in fungal community and diversity in strawberry rhizosphere soil after 12 years in the greenhouse. *Journal of Integrative Agriculture, 18*(3), 677–687. https://doi.org/10.1016/S2095-3119(18)62003-9.

Liu, H., Brettell, L. E., Qiu, Z., & Singh, B. K., (2020). Microbiome-mediated stress resistance in plants. *Trends Plant Sci., 25*, 733–743. https://doi.org/10.1016/j.tplants.2020.03.014.

Mehmood, S., Muneer, M. A., Tahir, M., Javed, M. T., Mahmood, T., Afridi, M. S., et al., (2021). Deciphering distinct biological control and growth promoting potential of multi-stress tolerant *Bacillus subtilis* PM32 for potato stem canker. *Physiol. Mol. Biol. Plants, 27*, 2101–2114. https://doi.org/10.1007/s12298-021-01067-2.

Mendes, R., Kruijt, M., De Bruijn, I., Dekkers, E., Van, D. V. M., Schneider, J. H. M., et al., (2011). Deciphering the rhizosphere microbiome for disease-suppressive bacteria. *Science (80), 332*, 1097–1100. https://doi.org/10.1126/science.1203980.

Micallef, S. A., Shiaris, M. P., & Colón-Carmona, A., (2009). Influence of *Arabidopsis thaliana* accessions on rhizobacterial communities and natural variation in root exudates. *J Exp Bot., 60*, 1729–1742. https://doi.org/10.1093/jxb/erp053.

Mueller, N. D., Gerber, J. S., Johnston, M., Ray, D. K., Ramankutty, N., & Foley, J. A., (2012). Closing yield gaps through nutrient and water management. *Nature, 490*, 254–257. https://doi.org/10.1038/nature11420.

Mueller, U., & Sachs, J. L., (2015). Engineering microbiomes to improve plant and animal health. *Trends Microbiol., 23*, 606–617. https://doi.org/10.1016/j.tim.2015.07.009.

Munns, R., James, R. A., Xu, B., Athman, A., Conn, S. J., Jordans, C., et al., (2012). Wheat grain yield on saline soils is improved by an ancestral Na+ transporter gene. *Nat. Biotechnol., 30*, 360–364. https://doi.org/10.1038/nbt.2120.

O'Neal, L., Vo, L., & Alexandre, G., (2020). Specific root exudate compounds sensed by dedicated chemoreceptors shape *Azospirillum brasilense* chemotaxis in the rhizosphere. *Appl. Environ. Microbiol., 86*. https://doi.org/10.1128/AEM.01026-20.

Panke-Buisse, K., Poole, A. C., Goodrich, J. K., Ley, R. E., & Kao-Kniffin, J., (2015). Selection on soil microbiomes reveals reproducible impacts on plant function. *ISME J., 9*, 980–989. https://doi.org/10.1038/ismej.2014.196.

Peiffer, J. A., Spor, A., Koren, O., Jin, Z., Tringe, S. G., Dangl, J. L., et al., (2013). Diversity and heritability of the maize rhizosphere microbiome under field conditions. *Proc. Natl. Acad. Sci., 110*, 6548–6553. https://doi.org/10.1073/pnas.1302837110.

Qiu, Z., Egidi, E., Liu, H., Kaur, S., & Singh, B. K., (2019). New frontiers in agriculture productivity: Optimized microbial inoculants and in situ microbiome engineering. *Biotechnol. Adv., 37,* 107371. https://doi.org/10.1016/j.biotechadv.2019.03.010.

Quiza, L., St-Arnaud, M., & Yergeau, E., (2015). Harnessing phytomicrobiome signaling for rhizosphere microbiome engineering. *Front Plant Sci., 6*. https://doi.org/10.3389/FPLS.2015.00507/FULL.

Raaijmakers, J. M., Paulitz, T. C., Steinberg, C., Alabouvette, C., & Moënne-Loccoz, Y., (2009). The rhizosphere: A playground and battlefield for soilborne pathogens and beneficial microorganisms. *Plant Soil., 321*, 341–361. https://doi.org/10.1007/s11104-008-9568-6.

Raaijmakers, J. M., Sluis, L. V. D., Bakker, P. A. H. M., Schippers, B., Koster, M., & Weisbeek, P. J., (1995). Utilization of heterologous siderophores and rhizosphere competence of fluorescent *Pseudomonas* spp. *Can. J. Microbiol., 41*, 126–135. https://doi.org/10.1139/m95-017.

Rahimi, S., Talebi, M., Baninasab, B., Gholami, M., Zarei, M., & Shariatmadari, H., (2020). The role of plant growth-promoting rhizobacteria (PGPR) in improving iron acquisition by altering physiological and molecular responses in quince seedlings. *Plant Physiol. Biochem., 155*, 406–415. https://doi.org/10.1016/j.plaphy.2020.07.045.

Raza, A., Razzaq, A., Mehmood, S. S., Zou, X., Zhang, X., Lv, Y., et al., (n.d.). *Impact of Climate Change on Crops Adaptation and Strategies to Tackle its Outcome: A Review*. MdpiCom. https://doi.org/10.3390/plants8020034.

Rodrigues, S. M., Demokritou, P., Dokoozlian, N., Hendren, C. O., Karn, B., Mauter, M. S., et al., (2017). Nanotechnology for sustainable food production: Promising opportunities and scientific challenges. *Environ. Sci Nano, 4*, 767–781. https://doi.org/10.1039/C6EN00573J.

Rodriguez, P. A., Rothballer, M., Chowdhury, S. P., Nussbaumer, T., Gutjahr, C., & Falter-Braun, P., (2019). Systems biology of plant-microbiome interactions. *Mol. Plant, 12*, 804–821. https://doi.org/10.1016/j.molp.2019.05.006.

Rosenberg, E., & Zilber-Rosenberg, I., (2013). *Role of Microorganisms in Adaptation, Development, and Evolution of Animals and Plants: The Hologenome Concept* (pp. 347–58). The Prokaryotes, Berlin, Heidelberg: Springer Berlin Heidelberg. https://doi.org/10.1007/978-3-642-30194-0_117.

Ryan, P. R., Dessaux, Y., Thomashow, L. S., & Weller, D. M., (2009). Rhizosphere engineering and management for sustainable agriculture. *Plant Soil., 321*, 363–383. https://doi.org/10.1007/S11104-009-0001-6.

Salles, J. F., Van, V. J. A., & Van, E. J. D., (2004). Multivariate analyses of *Burkholderia* species in soil: Effect of crop and land use history. *Appl. Environ. Microbiol., 70*, 4012–4020. https://doi.org/10.1128/AEM.70.7.4012-4020.2004.

Sankar, G. K., Sundaramoorthy, P., Nagarajan, M., & Lawrence, X. R., (2017). Role of organic amendments in sustainable agriculture. *Sustain. Agric. Towar. Food Secur.* (pp. 111–124). Singapore: Springer Singapore. https://doi.org/10.1007/978-981-10-6647-4_7.

Santos-Medellín, C., Edwards, J., Liechty, Z., Nguyen, B., & Sundaresan, V., (2017). Drought stress results in a compartment-specific restructuring of the rice root-associated microbiomes. *MBio., 8*. https://doi.org/10.1128/MBIO.00764-17/SUPPL_FILE/MBO004173388SF7.EPS.

Santoyo, G., Orozco-Mosqueda, M. D. C., & Govindappa, M., (2012). Mechanisms of biocontrol and plant growth-promoting activity in soil bacterial species of *Bacillus* and *Pseudomonas*: A review. *Biocontrol. Sci. Technol., 22*, 855–872. https://doi.org/10.1080/09583157.2012.694413.

Savary, S., Willocquet, L., Pethybridge, S. J., Esker, P., McRoberts, N., & Nelson, A., (2019). The global burden of pathogens and pests on major food crops. *Nat. Ecol. Evol., 3*, 430–439. https://doi.org/10.1038/s41559-018-0793-y.

Schipanski, M. E., Barbercheck, M., Douglas, M. R., Finney, D. M., Haider, K., Kaye, J. P., et al., (2014). A framework for evaluating ecosystem services provided by cover crops in agroecosystems. *Agric. Syst., 125*, 12–22. https://doi.org/10.1016/j.agsy.2013.11.004.

Singh, Y. P., Arora, S., Mishra, V. K., Dixit, H., & Gupta, R. K., (2019). Plant and soil responses to the combined application of organic amendments and inorganic fertilizers in degraded sodic soils of Indo-Gangetic plains. *Commun. Soil Sci. Plant Anal., 50*, 2640–2654. https://doi.org/10.1080/00103624.2019.1671446.

Syed, A. R. S. F., Singh, E., Pieterse, C. M. J., & Schenk, P. M., (2018). Emerging microbial biocontrol strategies for plant pathogens. *Plant Sci., 267*, 102–111. https://doi.org/10.1016/j.plantsci.2017.11.012.

Teixeira, P. J. P., Colaianni, N. R., Fitzpatrick, C. R., & Dangl, J. L., (2019). Beyond pathogens: Microbiota interactions with the plant immune system. *Curr. Opin. Microbiol., 49*, 7–17. https://doi.org/10.1016/j.mib.2019.08.003.

Tilman, D., Balzer, C., Hill, J., & Befort, B. L., (2011). Global food demand and the sustainable intensification of agriculture. *Proc. Natl. Acad. Sci U S A., 108*, 20260–20264. https://doi.org/10.1073/PNAS.1116437108/SUPPL_FILE/PNAS.201116437SI.PDF.

Timm, C. M., Carter, K. R., Carrell, A. A., Jun, S. R., Jawdy, S. S., Vélez, J. M., et al., (2018). Abiotic stresses shift belowground populus-associated bacteria toward a core stress microbiome. *MSystems., 3*. https://doi.org/10.1128/MSYSTEMS.00070-17/SUPPL_FILE/SYS001182164SF4.TIF.

Van, D. H. M. G. A., Bardgett, R. D., & Van, S. N. M., (2008). The unseen majority: Soil microbes as drivers of plant diversity and productivity in terrestrial ecosystems. *Ecol. Lett., 11*, 296–310. https://doi.org/10.1111/j.1461-0248.2007.01139.x.

Vannier, N., Agler, M., & Hacquard, S., (2019). Microbiota-mediated disease resistance in plants. *PLOS Pathog., 15*, e1007740. https://doi.org/10.1371/journal.ppat.1007740.

Vannier, N., Agler, M., & Hacquard, S., (2019). Microbiota-mediated disease resistance in plants. *PLOS Pathog., 15*, e1007740.

Wagg, C., Bender, S. F., Widmer, F., & Van, D. H. M. G. A., (2014). Soil biodiversity and soil community composition determine ecosystem multifunctionality. *Proc. Natl. Acad. Sci., 111*, 5266–5270. https://doi.org/10.1073/pnas.1320054111.

Wagg, C., Bender, S. F., Widmer, F., & Van, D. H. M. G. A., (2014). Soil biodiversity and soil community composition determine ecosystem multifunctionality. *Proc. Natl. Acad. Sci. U S A., 111*, 5266–5270. https://doi.org/10.1073/PNAS.1320054111.

York, L. M., Carminati, A., Mooney, S. J., Ritz, K., & Bennett, M. J., (2016). The holistic rhizosphere: Integrating zones, processes, and semantics in the soil influenced by roots. *J. Exp. Bot., 67*, 3629–3643. https://doi.org/10.1093/jxb/erw108.

Zhang, X., Huang, Y., Harvey, P. R., Ren, Y., Zhang, G., Zhou, H., et al., (2012). Enhancing plant disease suppression by *Burkholderia vietnamiensis* through chromosomal integration of *Bacillus subtilis* chitinase gene chi113. *Biotechnol. Lett., 34*, 287–293. https://doi.org/10.1007/S10529-011-0760-Z.

Zhao, L., Lu, L., Wang, A., Zhang, H., Huang, M., Wu, H., et al., (2020). Nano-biotechnology in agriculture: Use of nanomaterials to promote plant growth and stress tolerance. *J. Agric. Food Chem., 68*, 1935–1947. https://doi.org/10.1021/ACS.JAFC.9B06615.

Zuo, Y., & Zhang, F., (2011). Soil and crop management strategies to prevent iron deficiency in crops. *Plant Soil, 339*, 83–95. https://doi.org/10.1007/S11104-010-0566-0.

CHAPTER 9

Microbial Traits in Plant Stress Tolerance and Importance for Agriculture

MUHAMMAD AMMAR JAVED,[1] AROONA SALEEM,[1] AQSA HAFEEZ,[2] and BABER ALI[2]

[1]*Institute of Industrial Biotechnology, Government College University, Lahore, Pakistan*

[2]*Department of Plant Sciences, Quaid-i-Azam University, Islamabad, Pakistan*

ABSTRACT

In the past, the extravagant use of chemicals such as fertilizers and pesticides have created some ecological imbalances and environmental hazards across the globe. Moreover, ecological events like frost damage, heat waves, flooding, intense rains, and prolonged droughts are likely to increase in the near future due to disastrous climate changes. Therefore, a wide range of mitigation and alleviation strategies are required to subsist under such minacious weather conditions. Nowadays, plant growth-promoting rhizobacteria (PGPR) are gaining momentum as excellent biocontrol and biofertilizers to suppress pathogenic diseases and enhance the yield and growth of a wide range of food crops. However, isolation and screening of PGPR possessing the most effective traits are the most important steps for procuring benefits in fields. PGPR expresses different traits under biotic as well as abiotic stresses. They may produce antibiotics, bacteriocins, lytic enzymes, and siderophores, induce systemic resistance, or may cause signal interference to inhibit the chances of pathogen infections. Similarly, under abiotic stress conditions, they produce exopolysaccharides (EPS), and phytohormones, and exhibit ACC-deaminase activity, ROS-scavenging activity, phosphate

Advances in Plant Microbiome Research for Climate Resilient Agriculture: Toward Sustainable Farming. Ashwani Kumar (Ed.)

solubilization, and ion homeostasis to increase the crop yield. This chapter accentuates the beneficial traits of PGPR that should be kept into mind before their application in fields to gain maximum output as PGPR are highly specific to the soils and plant species.

9.1 INTRODUCTION

Plants encounter several environmental stresses that alter their growth and productivity. Such environmental stresses are split into two categories, i.e., biotic and abiotic stresses. Biotic stress in plants is attributed to living organisms including insects, nematodes, fungi, and bacteria. Biotic stress can deprive plants of their nutrients and cause infections which ultimately lead the host to death (Shanker & Shanker, 2016). On the other hand, water scarcity, water logging, salt concentrations, extreme temperatures, and heavy metal stresses are all consolidated into abiotic stresses (Saeed et al., 2022; Nawaz et al., 2022). Abiotic stress has noisome effects on plant growth, germination, seed quality, and crop yield (Zhang et al., 2022). Economic gains from agriculture are jeopardized by biotic and abiotic stress factors. On average, annually 17.2% to 30.0% of the food crops are scotched by weeds, pests, and pathogens (Savary et al., 2019). Similarly, 51% to 82% of food crops are devastated by some major abiotic stresses (Oshunsanya et al., 2019).

Chemical fertilizers, pesticides, herbicides, and fungicides have been used extensively to promote the growth of the plant and fortify them against pathogens. However, chemical treatments have their pros and cons and their drawbacks have limited their use. The use of chemicals can impart resistance to pests and is injurious to the health of consumers (Glick, 2014). They are not sustainable and most of the nitrogen and phosphorus of fertilizers do not reach the plant and are lost as a result of different processes (Adesemoye & Kloepper, 2009). So, there is a need to find alternatives to chemicals and microorganisms have emerged as a potential candidate that can easily replace chemical fertilizers as well as pesticides. They are environment friendly and act as biocontrol agents, biofertilizers, and phytostimulants (Babalola & Glick, 2012). Thus, researchers have directed their contemplation to the potential use of microorganisms in alleviating abiotic and biotic stresses. They promote plant growth, impart resistance to harsh environmental conditions, and prevent the chances of infection from pathogens. They also stabilize the soil and enhance traits of plants that are required to survive in stern abiotic and biotic stresses (Porter et al., 2020). This chapter will focus on the characteristics of microorganisms that alleviate the effects of abiotic and biotic stresses.

9.2 PLANT GROWTH-PROMOTING BACTERIA (PGPB)

The rhizosphere is home to a variety of microorganisms that form symbiotic relations with microorganisms and provides a medium for the exchange of signals and materials between plants and microorganisms (Qiao et al., 2017). Interaction in symbiosis can be synergistic, neutral, or negative depending upon the microbial species and the environmental conditions. Plant growth-promoting bacteria (PGPB) play a key role in the rhizosphere as they are involved in modifying plants, suppressing pathogens, and fixing nitrogen under biotic and abiotic stresses. PGPB aggrandizes the plant growth by imparting special traits, i.e., stress management, better nutrient uptake, and overall improvement in physiological activities of plants (Gamalero & Glick, 2022). PGPB have found to be involved in the heavy metal stress alleviation and betterment of nitrogen and phosphate conversions in soil. They also produce antagonistic metabolites to resist pathogens and boost immunity traits of crops under biotic stress. Under abiotic stresses, PGPB also alleviates drought, stress, heavy metal, nutrient, and chemical stress (Ayuso-Calles et al., 2021).

Understanding the mechanisms of plant-microorganism interaction is a complex process. However, the root secretions are found to be involved in attracting the microorganisms as these diverse organic compounds serve as a source of nutrients for microorganisms in the rhizosphere. Plants assemble and colonize the PGPB in the rhizosphere and provide them with nutrients (Amna et al., 2021). In return, microorganisms solubilize nitrogen and phosphate, mobilize nutrients, promote nutrients and water uptake, resist pathogens by competition and production of anti-metabolites, secrete plant-growth regulatory substances, and impart tolerance to plants against drought, salt, temperature, and heavy metals (Majeed et al., 2018) as shown in Figure 9.1. Several studies have been performed to determine the extent of benefits that can be exploited by the PGPB application. For instance, *Pseudomonas* sp. promotes plant growth, and improves water uptake in *Zea mays* under drought stress (Sandhya et al., 2010). Similarly, heavy metal, salinity, and drought stress in *Solanum tuberosum* are alleviated by *Bacillus pumilus* (Gururani et al., 2013). Shoot growth in *Avena sativa* is brought by *Actinobacter* sp. (Xun et al., 2015). The attack of *Fusarium oxysporum, Curvularia* sp., and *Alternaria* sp. is in *Oryza sativa* is resisted by *Paenibacillus baetica,* and *Pentoea vagans* (Verma et al., 2018). In this way, PGPB serves as a potential candidate against fertilizers and pesticides for the sustainability and improvement of agricultural yield in an ecologically stable manner.

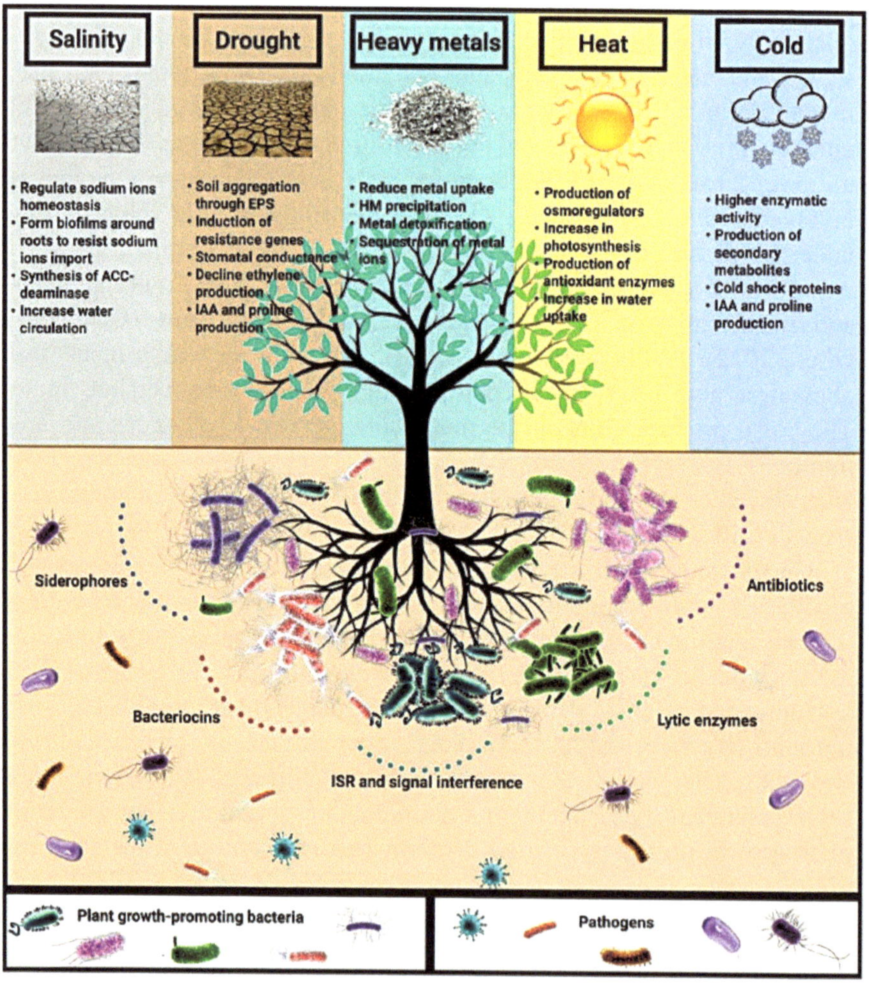

FIGURE 9.1 Role of plant growth-promoting bacteria in alleviation of abiotic and biotic stresses. The figure shows how PGPR help mitigate salinity, drought, heavy metals, heat and cold stress along with prevention of pathogens by producing siderophores, antibiotics, bacteriocins, lytic enzymes, induce systemic resistance and signal interference.

9.3 ALLEVIATION OF BIOTIC STRESSES

Over recent decades, pathogens have become a considerable menace to food production which has increased the use of agrochemical pesticides by farmers. The cost associated with the chemicals and the resistance they impart to the microorganisms has limited their use in agriculture. In this regard, microorganisms have occurred as a potential alternative to control phytopathogens, especially in the rhizosphere.

9.3.1 ANTIBIOSIS

One of the mechanisms used by microorganisms to prevent phytopathogens is antibiosis. Antibiosis is the production of certain low molecular weight compounds by beneficial microorganisms which act as antibiotics for phytopathogens (Reddy, 2013). Antibiosis is one of the most promising biocontrol mechanisms for impeding plant pathogens. Several PGPR produces antibiotics that are efficacious against a variety of fungal pathogens. Four classes of antibiotics, i.e., pyoluteorin (Plt), pyrrolnitrin (Prn), 2,4-diacetylphloroglucinol (DAPG), and phenazine-1-carboxylic acid (PCA) could be extracted from different strains of pseudomonads (Weller, 2007). The black rot of tobacco can be suppressed by DAPG produced by *Pseudomonas fluorescens*. Moreover, IAA, Prn, Plt, and siderophores (pseudobactin and pyochelin) are also produced by *P. fluorescens* which makes it the most efficacious plant growth-promoting rhizobacteria (PGPR) strain in biocontrol and plant growth (Weller, 2007; Weller et al., 2012). Various genes and traits in PGPR are found to be responsible for the production of antibiotics against pathogens. For instance, *P. chlororaphis* contains a marker gene, i.e., *phzf* for PGPR antibiotic PCA detection. PCA is a class of broad-spectrum antibiotics that suppresses *Fusarium* wilt and *Rhizoctonia* root rot. Similarly in *Pseudomonas* sp. 2,4-DAPG biosynthesis can be marked by the *phl*D gene (Wang et al., 2015). Pyoluteorin and pyrrolnitrin production in other *Pseudomonas* species, against *Rhizoctonia* and *Pythium*, is encoded by *plt*C and *prn*D genes (Loper et al., 2007). Certain *Bacillus* sp. also produce antibiotics like colistin, circulin, and polymyxin which are beneficial against gram-positive and gram-negative pathogens. For example, *B. cereus* produces kanosamine (aminoglycoside) and zwittermicin A (aminopylol) against oomycete pathogens in alfalfa crops (Mahmood et al., 2019).

9.3.2 BACTERIOCINS AND LYSIS ENZYMES

Bacteriocins are different from antibiotics in a way that bacteriocins are less lethal and only kill the bacteria which are closely related to the bacteriocin-producing bacteria (Riley & Wertz, 2002). Almost all bacteria can produce bacteriocins as their defense mechanisms. Some bacteriocins are a result of the recombination of different bacteriocins in gram-negative bacteria (Goh & Philip, 2015). Among bacteriocin-producing bacteria, *E. coli* is the most important bacteria that produce the representative bacteriocin, i.e., colicins. Similarly, *B. megaterium* produces megacins, *P. pyogenes* produces pyocins, *Serratia marcescens* produces marcescens, and *Enterobacter cloacae* produces cloacins (Cascales et al., 2007).

Several populations of PGPR secrete cell wall hydrolases, e.g., proteases, cellulases, glucanases, and chitinases which can resist the pathogens (Santoyo et al., 2021). Chitinase produced by *Serratia plumuthica* can impede the germination of the fungus *Botrytis cinerea* (Frankowski et al., 2001). Similarly, *Pseudomonas cepacia* produces glucanase which destroys the cell walls of *Sclerotium rolfsii, Pythium ultimum,* and *Rhizoctonia solani* (Fridlender et al., 1993). The cell wall of *Fusarium oxysporum* is lysed by the β-1,3-glucanase and chitinase produced by the *Streptomyces* sp. 385 and *Paenibacillus* sp. 300 (Singh et al., 1999). The mycelial growth of *B. cinerea* is halted by chitinase produced by *B. thuringiensis* UM96 (Kumar et al., 2012) and the growth of mycelia in *Phytophthora parasitica* is exalted by protease produced by *Paenibacillus* sp. B2 (Budi et al., 2000). These findings highlight the inhibitory potential of lysis enzymes produced by PGPR against multiple pathogens in the rhizosphere.

9.3.3 SIGNAL INTERFERENCE

Many pathogenic bacteria detect the accumulated levels of quorum-sensing molecules, also called autoinducers, to induce their virulence factors or pathogenicity. So, the mechanism of quorum sensing (QS) can be interfered by the degradation of autoinducers for purpose of the biocontrol of pathogens (Dutta et al., 2008). The lactone ring of N-acyl-homoserine lactones (AHLs) can be hydrolyzed by lactonases produced by *B. thuringiensis* (Dong et al., 2002) and AHL amide can be cleaved by acylases of *Ralstonia* sp. (Lin et al., 2003). Several species of PGPRs are involved in the production of AHL-degrading enzymes. Such PGPR includes *Microbacterium testaceum* StLB037 (Wang et al., 2010), *Mesorhizobium* sp. S5 (Mahmoudi et al., 2011),

Ochrobactrum sp. T63 (Mei et al., 2010), *Rhodococcus* sp. PI33 and LS31 (Park et al., 2006), *Streptomyces* sp. M664 (Park et al., 2005), *Rhodococcus erythripolis* W2 (Uroz et al., 2005), *Actinobacter* sp. C1010 (Kang et al., 2004), *Arthrobacter* sp. IBN110 (Park et al., 2003), *Comamonas* sp. D1 (Uroz et al., 2007), *Pseudomonas* sp. (Jayanna & Umesha, 2017), and *Bacillus* sp. (Han et al., 2010). The acylase produced by *P. aerogenosa* degrades the AHL as well as impedes the biofilm formation of *Ralstonia solanacearum* (Jayanna & Umesha, 2017). Moreover, AHL-QS of *Erwinia amylovora* and *Agrobacterium tumefaciens* can be quenched by *Bacillus simplex* and *Rhodococcus erythropolis* (Han et al., 2010; Hanano et al., 2014). In this way, the AHL-degrading property of PGPR can help biocontrol the phytopathogens in the rhizosphere to protect agricultural yields.

9.3.4 SIDEROPHORES

PGPB produces siderophores that reduce the bioavailability of iron for the pathogens by strongly binding with iron. Several species of PGPB produce siderophores that resist phytopathogens (Ahmed & Holmström, 2014). *Gaeumannomyces graminis* (barley and wheat growth deficiency) and *Fusarium oxysporum* (cause of potato wilt) can be suppressed by pyoverdines produced by *Pseudomonas* (Schippers et al., 1987; Voisard et al., 1989). *Fusarium* wilt of pepper can also be inhibited by the siderophores produced by *B. subtilis* (Yu et al., 2011). Several bacterial siderophores can also prevent the growth of fungal pathogens and protect agricultural crops against various diseases. Ferrioxamines produced by *Azadirachta indica* chelate the Fe^{3+} in the rhizosphere and hinder the growth of fungal pathogens (De Vleesschauwer & Höfte, 2009). *Fusarium* wilt of radish is suppressed by siderophores from *P. fluorescens* (Leeman et al., 1996). *Pseudomonas* siderophores also protect the wheat from *Pythium ultimum* (cause of *Pythium* root rot) (Loper, 1988). Fungal pathogens may also produce siderophores but bacterial siderophores are dominant in nature as they have more affinity for iron compared to fungal siderophores. That's how bacterial species outcompete the fungal pathogens in the rhizosphere and protect the crops.

9.3.5 INDUCED SYSTEMIC RESISTANCE (ISR)

The mechanism of resistance induced by the plants in response to the infection caused by the pathogen is called induced systemic resistance (ISR). Similarly, if the resistance is induced by the pathogen it is called systemic acquired resistance (SAR) (Sofy et al., 2019). ISR provides non-specific protection against several pathogens including nematodes, insects, viruses, bacteria, and fungi. ISR can be elucidated by various bacterial traits like the production of DAPG, salicylic acid (SA), siderophores, lipopolysaccharides (LPS), and flagella (Kannojia et al., 2019). ISR in *Arabidopsis thaliana* is elucidated by pseudobactin, LPS, and flagella of *P. putida* when applied exogenously to the roots (Meziane et al., 2005). Apart from these, ISR can also be stimulated by other bacterial traits. For instance, quorum-sensing molecules produced by *Serratia plymuthica* HRO-C48, *P. putida* IsoF, and *Serratia liquefaciens* MG1 (Schuhegger et al., 2006; Pang et al., 2009), volatile organic compounds (VOCs) produced by *Bacillus subtilis* (2R,3R-butanediol), dimethyl sulfide (DMDS) by *B. cereus* C1L, and branched-chain alcohols by *B. amyloliquefaciens* IN937a (Ryu et al., 2004; Farag et al., 2006). PGPR possesses a great potential of inducing systemic resistance and its efficacy depends upon the traits specific to a strain. In some cases, two bacterial populations work synergistically to induce systemic resistance and suppress pathogenic diseases. ISR expression of *Macrophomina phaseolina* is enhanced significantly by the application of microbial consortium of *B. subtilis* CRN-16 and *P. putida* CRN-09 in mung beans compared to the use of a single strain (Sharma et al., 2018). Similarly, *Microbacterium* sp., *Stentrophormonas* sp., and *Xanthomonas* sp. are ineffective separately but when combined in the microbial consortium, they can effectively induce ISR against *Hyaloperonospora arabidopsidis* (Berendsen et al., 2018).

9.4 ALLEVIATION OF ABIOTIC STRESSES

Several abiotic stresses could perturb the productivity of agricultural fields. Abiotic factors, i.e., salt, drought, heat, toxin, and cold are interlinked with each other and can simultaneously decrease the plant growth and ultimately agricultural yield (Hussain et al., 2022). For instance, salt stress limits the uptake of minerals (Ca^{+2} and K^{+}) and water which ultimately results in less cytosolic metabolism, and decreased cell expansion. Drought stress reduces the plant shoot growth and metabolic demands. High temperature declines the efficacy of photosynthesis, seed germination, and yield. Similarly, cold stress alters the signal transduction pathways and cellular functions of plants (Cramer et al., 2011). Plants develop symbiotic relations with microorganisms to mitigate these stresses and increase crop yield. Following mechanisms are adopted by microorganisms for alleviating abiotic stresses.

9.4.1 ACC DEAMINASE ACTIVITY

Ethylene (plant hormone) is produced by the plants to modulate diverse mechanisms including root elongation, nodules, and roots development, senescence, fruit ripening, seed germination, etc. However, ethylene can also be produced under stress conditions, i.e., drought, salt, temperature, heavy metals, phytopathogens, and flooding and such ethylene is termed 'stress ethylene' (Glick, 2014). PGPRs possess an enzyme 1-amino-cyclopropane-1-carboxylate (ACC) deaminase which protects the plant from stress ethylene and promotes plant growth. The cleavage of ACC to α-ketobutyrate and ammonia is catalyzed by PGPR-mediated ACC deaminase which leads to a reduction in the concentration of stress ethylene. PGPRs act as reservoirs of ACC in the rhizosphere to reduce ethylene and enhance root growth. Therefore, due to extensive root growth, plants can tolerate various stress conditions more efficaciously in presence of ACC deaminase-containing PGPR (Glick et al., 2007). A variety of PGPRs is present in the rhizosphere of plants. For example, *Bacillus firmis* and *B. pumilus* enhanced the ROS scavenging activity of enzymes and increased the tolerance of *Solanum tuberosum* towards heavy metals, drought, and salinity stress (Gururani et al., 2013). In the rhizosphere of *Triticum aestivum*, PGPRs (*Ochrobactrum pseudogrignonense* and *B. safensis*) increased the accumulation of anti-oxidants (ascorbate and carotenoids) under drought stress (Chakraborty et al., 2013). *Paenibacillus polymyxa* and *Bacillus thuringiensis* inoculants increased the survivorship up to five-folds of *T. aestivum* and enhanced their biomass under severe drought conditions (Timmusk et al., 2014). The increase in the number of buds and flowers, higher chlorophyll content, and higher fresh and dry biomass are all attributed to the ACC deaminase activity of *Pseudomonas migulae* and *P. fluorescens* in *S. lycopersicum* plants under high salt stress (Ali et al., 2014). Similarly, under cold stress, *Burkholderia phytofirmans* increase the biomass and root growth of *Vitis vinifera* (Ait Barka et al., 2006). ACC deaminase activity of PGPRs is also seen in the alleviation of heavy metal stress in plants. *Pseudomonas taiwanensis* and *P. grimonti* elevated the growth and yield of *Cicer arietinum* in arsenic-contaminated soils (Shagol et al., 2014). The cumulative damage caused by zinc in *Zea mays* is prevented by ACC-deaminase activity of *Proteus mirabilis* (Islam et al., 2014). The growth of *Helianthus annuus* is promoted by PGPRs (*Klebsiella* sp., *Serratia* sp., and *Enterobacter* sp.) under high concentrations of lead, arsenic, and copper (Carlos et al., 2016).

9.4.2 EXOPOLYSACCHARIDES (EPS)

Exopolysaccharides (EPS) or extracellular polymeric substances are natural blends of high molecular weight polymers that are secreted by PGPRs to form biofilms and adhere to plant root surfaces. Under abiotic stress conditions, bacteria release EPS as a defense strategy to protect themselves against environmental variations. Therefore, abiotic stresses on plants can also be alleviated by the use of EPS-producing PGPR. Several studies have been performed to understand the mechanism of stress alleviation by EPS-producing bacteria.

Under salt stress, *Bacillus insolitus, B. polymyxa, B. circulans,* and *Aeromonas hydrophila/caviae* cover a high proportion of soil sheaths which restricts the uptake of Na^+ in wheat plants (Ashraf et al., 2004). EPS of *Serratia proteamaculans* and *B. subtilis* binds the free Na^+ and restricts the NA^+ influx in soybean plants (Han & Lee, 2005). In an experiment performed by Arora et al. (2010), Cyanobacteria adsorbed the Na^+ from the solution reducing its availability for rice, maize, and wheat plants. *Staphylococcus saprophyticus* and *Oceanobacillus profundus* increase the root adherence to soil and chelate Na^+ in the rhizosphere of *Lens esculenta* (Qurashi & Sabri, 2012). Biofilm formation and soil aggregation are enhanced in Chickpea by *Planococcus rifietoensis* and *Halomonas variabilis* (Qurashi & Sabri, 2011). *B. amyloliquefaciens* improves the Na^+ homeostasis in maize and biofilm stability in barley (Kasim et al., 2016).

Under drought stress, EPS acts as a conductor and reservoir or water for plant roots. EPS retains water and improves soil structure by aggrandizing the biofilm formation and bacterial colonization. It enhances the survival rate of plants by providing them with the time to bring metabolic adjustments under drought stress. For instance, *B. polymyxa* enhances the root adhering soil to root tissue ratio (RAS/RT) in wheat crops under drought stress (Bezzate et al., 2000). In another study, RAS was also enhanced by *Pantoea agglomerans* in wheat (Amellal et al., 1998). Soil aggregate stability is enhanced in sunflower (by *P. putida*) (Sandhya et al., 2009), maize (by *Brevibacillus brevis*, and *Pseudomonas aeruginosa*) (Putrie et al., 2013), and rice (by *Azotobacter* sp.) (Sivapriya & Priya, 2017) under drought stress. Maize can be protected from drought by increasing the water and moisture content of soil using *Alcaligenes faecalis, P. aeruginosa,* and *Proteus penneri* (Naseem & Bano, 2014). According to Zheng et al. (2018), polysaccharides from EPS-containing PGPR modify the soil structure and reduce the water evaporation and hydraulic conductivity to provide more water to plants. Similarly, water evaporation was limited at pore throats to reduce water loss in soil by the *Sinorhizobium meliloti*-derived EPS (Deng et al., 2015).

9.4.3 PHYTOHORMONES

Several free-living, endophytic, and symbiotic microorganisms can produce low amounts of various phytohormones (auxins, cytokinins, gibberellins, ABA) to promote the growth of plants and withstand stress efficaciously (Hashem et al., 2016). Auxins especially indole acetic acid (IAA) enhance the root growth and number of root branches in the rhizosphere even under stress conditions. This increase in the root network improves the nutrients and water uptake and imparts resistance against abiotic stresses (Wahab et al., 2022). For instance, salt-resistance in barley is stimulated by IAA produced by *Ensifer garamanticus* and *Curtobacterium flaccumfaciens* (Cardinale et al., 2015). Similarly, wheat production (*B. licheniformis*) (Singh & Jha, 2016), wheat biomass (*Arthrobacter* sp. and *B. subtilis*) (Upadhyay et al., 2012), and seed germination (*Enterobacter* sp.) (Sorty et al., 2016) was enhanced by IAA to mitigate salt stress. The root and shoot biomass of maize was increased by *Pseudomonas* sp. in high salt concentrations (Mishra et al., 2017). Clover biomass was seen to be enhanced in drought stress when seeds were treated with *B. megaterium* and *P. putida* (Marulanda et al., 2009). Similarly, in nutrient deficient soil IAA-producing *Serratia* sp. boosted the grain yield of chickpea (Zaheer et al., 2016). Microbial-derived phytohormones also inflate the phytoextraction of metals. The root network of *Brassica juncea* was amplified by IAA production activities performed by *A. xylosoxidans* to enhance copper phytoextraction (Ma et al., 2009). Nickle stress in *B. juncea* L. (Zaidi et al., 2006) and *Vinca rosea* (Khan et al., 2017) is mitigated by IAA-producing *B. megaterium*. The accumulation of ascorbate peroxidase (APX), peroxidase (POD), catalase (CAT), and superoxide dismutase (SOD) was also aggravated in *B. megaterium* treated plants compared to untreated ones.

9.4.4 OTHER TRAITS

Abiotic stresses like drought could lead to the misdirection of an electron during photosynthesis, resulting in the formation of reactive oxygen species (ROS), which induces oxidative stress. So, ROS-degrading bacteria could mitigate oxidative stress from plants. For instance, ROS are degraded by abscisic acid (ABA), accumulated as a result of antioxidants and cytokinins produced by *Azotobacter chroococcum* (Grover et al., 2011). *Azotobacter* and *Azospirillum* also increased the activities of antioxidant enzymes, i.e., CAT, peroxidase (POX), and SOD (Karthikeyan et al., 2012). Under very low temperatures, *Pseudomonas frederiksbergensis* and *P. vancouverensis* code for anti-chilling proteins in *Solanum lycopersicum*. It also protects from membrane damage and reduces the level of ROS (Subramanian et al., 2015). Under drought stress, *Azospirillum brasilense* improves the water status of *Triticum aestivum* resulting in better uptake of minerals (Choudhary et al., 2016). Proline is produced by plants under stress to improve growth by acting as a reactive oxygen scavenger. Certain legumes like *Phaseolus vulgaris* and *Glycine max* accumulate proline under prolonged drought stress and direct relation is evident between drought tolerance and proline (Sharma et al., 2013). Under salt stress, *Pseudomonas* and *Rhizobium* stimulate proline production in *Zea mays* to maintain relative water content (Ali, Wang, Saleem, Hafeez et al., 2022).

Certain microorganisms can convert the insoluble phosphate compounds (inorganic and organic) into the form accessible to the plants. PGPR could solubilize the rock phosphate, hydroxyapatite, dicalcium phosphate, and tricalcium phosphate. PGPR contains phosphatase enzymes like phosphohydrolases to convert the organic P into the inorganic P for plant nutrition (Ahmad et al., 2022). Similarly, to utilize the free iron present in the soil, bacteria produce siderophores (iron-chelating compounds). They chelate the ferric ion with high affinity and solubilize it which increases the uptake of iron in plants (Zainab et al., 2021).

Moreover, PGPR could also alter the plant ion homeostasis to improve survival in salinity stress. A significant deficiency in K^+ ions take place under salt stress due to the accumulation of Na^+ ions (Ali, Wang, Saleem, Azeem et al., 2022). PGPR could increase the K^+ concentration and K^+/Na^+ ratio in the plants grown in saline soil. A deficit in K^+ could alter the stomatal openings, changing the leaf water status, which could also be alleviated by PGPR (Mehmood et al., 2021). For instance, salt stress in *Arabidopsis thaliana* could be alleviated by the regulation of K^+ transporter HKT1 by *Bacillus subtilis*. *B. subtilis* emit volatile compounds that increase affinity for K^+ transporters and mitigate salt stress (Zhang et al., 2008).

9.5 CONCLUSION

The disastrous effects of indiscriminate and extended use of chemicals and environmental stress factors have triggered the adoption of alternative methods for sustainable agriculture. Several microbial species exhibit traits for alleviation of biotic and abiotic stresses. The rhizosphere, a hub of microbial diversity, serves as a source of beneficial microorganisms that can be scrutinized and used for their benefit in agricultural fields. A variety of PGPR has a positive influence directly or indirectly on plant growth under severe biotic and abiotic stress conditions. They can protect the plants from pathogens attacks by the production of antibiotics, bacteriocins, lysis enzymes, signal interference, siderophores production, and induction of systemic resistance. Apart from biotic stresses, PGPR can also mitigate abiotic stresses like drought, extreme temperature, salinity, nutrient scarcity, and floods by the production of phytohormones, EPS, and induction of ACC-deaminase activity, ROS degradation, and solubilization of phosphates. All such traits could be assessed and examined while developing the PGPR strain for agricultural application. The choice of a suitable strain is the cornerstone of inoculum development. Moreover, new associations that could be formed as a result of inoculum application should also be investigated prior to use in agriculture fields.

KEYWORDS

- **abiotic stresses**
- **ACC-deaminase activity**
- **exopolysaccharides**
- **phosphate solubilization**
- **plant growth-promoting rhizobacteria**
- **ROS-scavenging activity**
- **siderophores**

REFERENCES

Adesemoye, A. O., & Kloepper, J. W., (2009). Plant–microbes interactions in enhanced fertilizer-use efficiency. *Appl. Microbiol. Biotechnol., 85*(1), 1–12.

Ahmad, M., Ishaq, M., Shah, W. A., Adnan, M., Fahad, S., Saleem, M. H., ... & Hashem, M. (2022). Managing phosphorus availability from organic and inorganic sources for optimum wheat production in calcareous soils. *Sustainability, 14*(13), 7669.

Ahmed, E., & Holmström, S. J. M., (2014). Siderophores in environmental research: Roles and applications. *Microb. Biotechnol., 7*(3), 196–208.

Ait, B. E., Nowak, J., & Clément, C., (2006). Enhancement of chilling resistance of inoculated grapevine plantlets with a plant growth-promoting rhizobacterium, *Burkholderia phytofirmans* strain PsJN. *Appl. Environ. Microbiol., 72*(11), 7246–7252.

Ali, B., Wang, X., Saleem, M. H., Azeem, M. A., Afridi, M. S., Nadeem, M., Ghazal, M., et al., (2022). Bacillus mycoides PM35 reinforces photosynthetic efficiency, antioxidant defense, expression of stress-responsive genes, and ameliorates the effects of salinity stress in maize. *Life, 12*(2), 219.

Ali, B., Wang, X., Saleem, M. H., Hafeez, A., Afridi, M. S., Khan, S., Ullah, I., et al., (2022). PGPR-mediated salt tolerance in maize by modulating plant physiology, antioxidant defense, compatible solutes accumulation and bio-surfactant producing genes. *Plants, 11*(3), 345.

Ali, S., Charles, T. C., & Glick, B. R., (2014). Amelioration of high salinity stress damage by plant growth-promoting bacterial endophytes that contain ACC deaminase. *Plant Physiol. Biochem.; PPB., 80,* 160–167. https://doi.org/10.1016/j.plaphy.2014.04.003.

Amellal, N., Burtin, G., Bartoli, F., & Heulin, T., (1998). Colonization of wheat roots by an exopolysaccharide-producing *Pantoea agglomerans* strain and its effect on rhizosphere soil aggregation. *Appl. Environ. Microbiol., 64*(10), 3740–3747.

Amna, Ali, B., Azeem, M. A., Qayyum, A., Mustafa, G., Ahmad, M. A., Javed, M. T., & Chaudhary, H. J., (2021). Bio-fabricated silver nanoparticles: A sustainable approach for augmentation of plant growth and pathogen control. In: *Sustain Agric. Rev.* (Vol. 53, pp. 345–371). [place unknown]: Springer.

Arora, M., Kaushik, A., Rani, N., & Kaushik, C. P., (2010). Effect of cyanobacterial exopolysaccharides on salt stress alleviation and seed germination. *J. Environ. Biol., 31*(5), 701–704.

Ashraf, M., Hasnain, S., Berge, O., & Mahmood, T., (2004). Inoculating wheat seedlings with exopolysaccharide-producing bacteria restricts sodium uptake and stimulates plant growth under salt stress. *Biol. Fertil. Soils, 40*(3), 157–162.

Ayuso-Calles, M., Flores-Félix, J. D., & Rivas, R., (2021). Overview of the role of rhizobacteria in plant salt stress tolerance. *Agronomy, 11*(9), 1759.

Babalola, O. O., & Glick, B. R., (2012). The use of microbial inoculants in African agriculture: Current practice and future prospects. *J. Food Agric. Env., 10*(3, 4), 540–549.

Berendsen, R. L., Vismans, G., Yu, K., Song, Y., De Jonge, R., Burgman, W. P., Burmølle, M., et al., (2018). Disease-induced assemblage of a plant-beneficial bacterial consortium. *ISME J., 12*(6), 1496–1507.

Bezzate, S., Aymerich, S., Chambert, R., Czarnes, S., Berge, O., & Heulin, T., (2000). Disruption of the *Paenibacillus polymyxa* levansucrase gene impairs its ability to aggregate soil in the wheat rhizosphere. *Environ. Microbiol., 2*(3), 333–342.

Budi, S. W., Van, T. D., Arnould, C., Dumas-Gaudot, E., Gianinazzi-Pearson, V., & Gianinazzi, S., (2000). Hydrolytic enzyme activity of *Paenibacillus* sp. strain B2 and effects of the antagonistic bacterium on cell integrity of two soil-borne pathogenic fungi. *Appl. Soil Ecol., 15*(2), 191–199.

Cardinale, M., Ratering, S., Suarez, C., Montoya, A. M. Z., Geissler-Plaum, R., & Schnell, S., (2015). Paradox of plant growth promotion potential of rhizobacteria and their actual promotion effect on growth of barley (*Hordeum vulgare* L.) under salt stress. *Microbiol. Res., 181,* 22–32.

Carlos, M. H. J., Stefani, P. V. Y., Janette, A. M., Melani, M. S. S., & Gabriela, P. O., (2016). Assessing the effects of heavy metals in ACC deaminase and IAA production on plant growth-promoting bacteria. *Microbiol. Res., 188,* 53–61.

Cascales, E., Buchanan, S. K., Duché, D., Kleanthous, C., Lloubes, R., Postle, K., Riley, M., et al., (2007). Colicin biology. *Microbiol. Mol. Biol. Rev., 71*(1), 158–229.

Chakraborty, U., Chakraborty, B. N., Chakraborty, A. P., & Dey, P. L., (2013). Water stress amelioration and plant growth promotion in wheat plants by osmotic stress tolerant bacteria. *World J. Microbiol. Biotechnol., 29*(5), 789–803.

Choudhary, D. K., Kasotia, A., Jain, S., Vaishnav, A., Kumari, S., Sharma, K. P., & Varma, A., (2016). Bacterial-mediated tolerance and resistance to plants under abiotic and biotic stresses. *J. Plant Growth Regul., 35*(1), 276–300. https://doi.org/10.1007/s00344-015-9521-x.

Cramer, G. R., Urano, K., Delrot, S., Pezzotti, M., & Shinozaki, K., (2011). Effects of abiotic stress on plants: A systems biology perspective. *BMC Plant Biol., 11*(1), 1–14.

De Vleesschauwer, D., & Höfte, M. (2009). Rhizobacteria-induced systemic resistance. *Advances in Botanical Research, 51*, 223–281.

Deng, J., Orner, E. P., Chau, J. F., Anderson, E. M., Kadilak, A. L., Rubinstein, R. L., Bouchillon, G. M., et al., (2015). Synergistic effects of soil microstructure and bacterial EPS on drying rate in emulated soil micromodels. *Soil Biol. Biochem. [Internet]. 83*, 116–124. https://doi.org/https://doi.org/10.1016/j.soilbio.2014.12.006.

Dong, Y. H., Gusti, A. R., Zhang, Q., Xu, J. L., & Zhang, L. H., (2002). Identification of quorum-quenching N-acyl homoserine lactonases from *Bacillus species. Appl. Environ. Microbiol., 68*(4), 1754–1759.

Dutta, S., Mishra, A. K., & Kumar, B. S. D., (2008). Induction of systemic resistance against fusarial wilt in pigeon pea through interaction of plant growth promoting rhizobacteria and rhizobia. *Soil Biol. Biochem., 40*(2), 452–461.

Farag, M. A., Ryu, C. M., Sumner, L. W., & Paré, P. W., (2006). GC–MS SPME profiling of rhizobacterial volatiles reveals prospective inducers of growth promotion and induced systemic resistance in plants. *Phytochemistry, 67*(20), 2262–2268.

Frankowski, J., Lorito, M., Scala, F., Schmid, R., Berg, G., & Bahl, H., (2001). Purification and properties of two chitinolytic enzymes of *Serratia plymuthica* HRO-C48. *Arch Microbiol., 176*(6), 421–426.

Fridlender, M., Inbar, J., & Chet, I., (1993). Biological control of soilborne plant pathogens by a β-1, 3 glucanase-producing *Pseudomonas cepacia. Soil Biol Biochem., 25*(9), 1211–1221.

Gamalero, E., & Glick, B. R., (2022). Recent advances in bacterial amelioration of plant drought and salt stress. *Biology (Basel), 11*(3), 437.

Glick, B. R., (2014). Bacteria with ACC deaminase can promote plant growth and help to feed the world. *Microbiol Res., 169*(1), 30–39.

Glick, B. R., Cheng, Z., Czarny, J., & Duan, J., (2007). Promotion of plant growth by ACC deaminase-producing soil bacteria. *New Perspect. Approaches Plant Growth-Promoting Rhizobacteria Res.*, 329–339.

Goh, H. F., & Philip, K., (2015). Purification and characterization of bacteriocin produced by *Weissella confusa* A3 of dairy origin. *PLoS One, 10*(10), e0140434.

Grover, M., Ali, S. Z., Sandhya, V., Rasul, A., & Venkateswarlu, B., (2011). Role of microorganisms in adaptation of agriculture crops to abiotic stresses. *World J. Microbiol Biotechnol., 27*(5), 1231–1240. https://doi.org/10.1007/s11274-010-0572-7.

Gururani, M. A., Upadhyaya, C. P., Baskar, V., Venkatesh, J., Nookaraju, A., & Park, S. W., (2013). Plant growth-promoting rhizobacteria enhance abiotic stress tolerance in *Solanum tuberosum* through inducing changes in the expression of ROS-scavenging enzymes and improved photosynthetic performance. *J. Plant Growth Regul., 32*(2), 245–258.

Han, H. S., & Lee, K. D., (2005). Plant growth promoting rhizobacteria effect on antioxidant status, photosynthesis, mineral uptake, and growth of lettuce under soil salinity. *Res. J. Agric. Biol. Sci., 1*(3), 210–215.

Han, Y., Chen, F., Li, N., Zhu, B., & Li, X., (2010). *Bacillus marcorestinctum* sp. Nov., a novel soil acylhomoserine lactone quorum-sensing signal quenching bacterium. *Int. J. Mol. Sci., 11*(2), 507–520.

Hanano, A., Harba, M., Al-Ali, M., & Ammouneh, H., (2014). Silencing of *Erwinia amylovora* sy69 AHL-quorum sensing by a *Bacillus* simplex AHL-inducible *aiiA* gene encoding a zinc-dependent N-acyl-homoserine lactonase. *Plant Pathol., 63*(4), 773–783.

Hashem, A., Abd_Allah, E. F., Alqarawi, A. A., Al-Huqail, A. A., Wirth, S., & Egamberdieva, D., (2016). The interaction between arbuscular mycorrhizal fungi and endophytic bacteria enhances plant growth of *Acacia gerrardii* under salt stress. *Front Microbiol., 7*, 1089.

Hussain, S., Rasheed, M., Hamzah, S. M., Ahmed, Z., Hafeez, A., Jilani, G., Alamri, S., et al., (2022). Salt tolerance in maize with melatonin priming to achieve sustainability in yield in salt affected soils. *Pakistan J. Bot.* https://doi.org/10.30848/PJB2023-1(27).

Islam, F., Yasmeen, T., Riaz, M., Arif, M. S., Ali, S., & Raza, S. H., (2014). Proteus mirabilis alleviates zinc toxicity by preventing oxidative stress in maize (*Zea mays*) plants. *Ecotoxicol. Environ. Saf., 110*, 143–152.

Jayanna, S. K., & Umesha, S., (2017). Quorum quenching activity of rhizosphere bacteria against *Ralstonia solanacearum*. *Rhizosphere, 4*, 22–24.

Kang, B. R., Lee, J. H., Ko, S. J., Lee, Y. H., Cha, J. S., Cho, B. H., & Kim, Y. C., (2004). Degradation of acyl-homoserine lactone molecules by *Acinetobacter* sp. strain C1010. *Can J. Microbiol., 50*(11), 935–941.

Kannojia, P., Choudhary, K. K., Srivastava, A. K., & Singh, A. K., (2019). PGPR bioelicitors: Induced systemic resistance (ISR) and proteomic perspective on biocontrol. In: *PGPR Amelior Sustain Agric.* (pp. 67–84). [place unknown]. Elsevier.

Karthikeyan, B., Joe, M. M., Islam, M., & Sa, T., (2012). ACC deaminase containing diazotrophic endophytic bacteria ameliorate salt stress in *Catharanthus roseus* through reduced ethylene levels and induction of antioxidative defense systems. *Symbiosis, 56*(2), 77–86.

Kasim, W. A., Gaafar, R. M., Abou-Ali, R. M., Omar, M. N., & Hewait, H. M., (2016). Effect of biofilm forming plant growth promoting rhizobacteria on salinity tolerance in barley. *Ann. Agric. Sci., 61*(2), 217–227.

Khan, W. U., Ahmad, S. R., Yasin, N. A., Ali, A., Ahmad, A., & Akram, W., (2017). Application of *Bacillus megaterium* MCR-8 improved phytoextraction and stress alleviation of nickel in *Vinca rosea. Int. J. Phytoremediation. 19*(9), 813–824.

Kumar, P., Dubey, R. C., & Maheshwari, D. K., (2012). Bacillus strains isolated from rhizosphere showed plant growth promoting and antagonistic activity against phytopathogens. *Microbiol. Res., 167*(8), 493–499.

Leeman, M., Den, O. F. M., Van, P. J. A., Dirkx, F. P. M., Steijl, H., Bakker, P., & Schippers, B., (1996). Iron availability affects induction of systemic resistance to *Fusarium* wilt of radish by *Pseudomonas fluorescens*. *Phytopathology, 86*(2), 149–155.

Lin, Y., Xu, J., Hu, J., Wang, L., Ong, S. L., Leadbetter, J. R., & Zhang, L., (2003). Acyl-homoserine lactone acylase from *Ralstonia* strain XJ12B represents a novel and potent class of quorum-quenching enzymes. *Mol. Microbiol., 47*(3), 849–860.

Loper, J. E., (1988). Role of fluorescent siderophore production in biological control of *Pythium ultimum* by a *Pseudomonas fluorescens* strain. *Phytopathology, 78*(2), 166–172.

Loper, J. E., Kobayashi, D. Y., & Paulsen, I. T., (2007). The genomic sequence of *Pseudomonas fluorescens* Pf-5: Insights into biological control. *Phytopathology, 97*(2), 233–238.

Ma, Y., Rajkumar, M., & Freitas, H., (2009). Inoculation of plant growth promoting bacterium *Achromobacter xylosoxidans* strain Ax10 for the improvement of copper phytoextraction by *Brassica juncea. J. Environ. Manage, 90*(2), 831–837.

Mahmood, I., Rizvi, R., Sumbul, A., & Ansari, R. A., (2019). Plant health under biotic stress. *Plant Heal Under Biot. Stress*,177–188. https://doi.org/10.1007/978-981-13-6040-4.

Mahmoudi, E., Tabatabaei, B. E. S., & Venturi, V., (2011). Virulence attenuation of *Pectobacterium carotovorum* using N-acyl-homoserine lactone degrading bacteria isolated from potato rhizosphere. *Plant Pathol. J., 27*(3), 242–248.

Majeed, A., Muhammad, Z., & Ahmad, H., (2018). Plant growth promoting bacteria: Role in soil improvement, abiotic and biotic stress management of crops. *Plant Cell Rep. [Internet]. 37*(12), 1599–1609. https://doi.org/10.1007/s00299-018-2341-2.

Marulanda, A., Barea, J. M., & Azcón, R., (2009). Stimulation of plant growth and drought tolerance by native microorganisms (AM fungi and bacteria) from dry environments: Mechanisms related to bacterial effectiveness. *J. Plant Growth Regul., 28*(2), 115–124.

Mehmood, S., Khatoon, Z., Amna, Ahmad, I., Muneer, M. A., Kamran, M. A., Ali, J., Ali, B., Chaudhary, H. J., & Munis, M. F. H., (2021). *Bacillus* sp. PM31 harboring various plant growth-promoting activities regulates *Fusarium* dry rot and wilt tolerance in potato. *Arch Agron. Soil Sci.*, 1–15.

Mei, G. Y., Yan, X. X., Turak, A., Luo, Z. Q., & Zhang, L. Q., (2010). AidH, an alpha/beta-hydrolase fold family member from an *Ochrobactrum* sp. strain, is a novel N-acylhomoserine lactonase. *Appl. Environ. Microbiol., 76*(15), 4933–4942.

Meziane, H., Van, D. S. I., Van, L. L. C., Höfte, M., & Bakker, P. A. H. M., (2005). Determinants of *Pseudomonas putida* WCS358 involved in inducing systemic resistance in plants. *Mol. Plant Pathol., 6*(2), 177–185.

Mishra, S. K., Khan, M. H., Misra, S., Dixit, V. K., Khare, P., Srivastava, S., & Chauhan, P. S., (2017). Characterization of *Pseudomonas* spp. and *Ochrobactrum* sp. isolated from volcanic soil. *Antonie Van Leeuwenhoek, 110*(2), 253–270.

Naseem, H., & Bano, A., (2014). Role of plant growth-promoting rhizobacteria and their exopolysaccharide in drought tolerance of maize. *J. Plant Interact. [Internet], 9*(1), 689–701. https://doi.org/10.1080/17429145.2014.902125.

Nawaz, H., Ali, A., Saleem, M. H., Ameer, A., Hafeez, A., Alharbi, K., Ezzat, A., et al., (2022). Comparative effectiveness of EDTA and citric acid assisted phytoremediation of Ni contaminated soil by using canola (*Brassica napus*). *Brazilian J. Biol. 82.*

Oshunsanya, S. O., Nwosu, N. J., & Li, Y., (2019). Abiotic stress in agricultural crops under climatic conditions. In: *Sustain. Agric. for Environ. Manag.* (pp. 71–100). [place unknown]: Springer.

Pang, Y., Liu, X., Ma, Y., Chernin, L., Berg, G., & Gao, K., (2009). Induction of systemic resistance, root colonization and biocontrol activities of the rhizospheric strain of *Serratia plymuthica* are dependent on N-acyl homoserine lactones. *Eur. J. Plant Pathol., 124*(2), 261–268.

Park, S. Y., Hwang, B. J., Shin, M. H., Kim, J. A., Kim, H. K., & Lee, J. K., (2006). N-acylhomoserine lactonase producing *Rhodococcus* spp. with different AHL-degrading activities. *FEMS Microbiol Lett., 261*(1), 102–108.

Park, S. Y., Kang, H. O., Jang, H. S., Lee, J. K., Koo, B. T., & Yum, D. Y., (2005). Identification of extracellular N-acylhomoserine lactone acylase from a *Streptomyces* sp. and its application to quorum quenching. *Appl. Environ. Microbiol., 71*(5), 2632–2641.

Park, S. Y., Lee, S. J., Oh, T. K., Oh, J. W., Koo, B. T., Yum, D. Y., & Lee, J. K., (2003). AhlD, an N-acylhomoserine lactonase in *Arthrobacter* sp., and predicted homologues in other bacteria. *Microbiology, 149*(6), 1541–1550.

Porter, S. S., Bantay, R., Friel, C. A., Garoutte, A., Gdanetz, K., Ibarreta, K., Moore, B. M., Shetty, P., Siler, E., & Friesen, M. L., (2020). Beneficial microbes ameliorate abiotic and biotic sources of stress on plants. *Funct Ecol., 34*(10), 2075–2086.

Putrie, R. F. W., Wahyudi, A., Nawansih, A. A., & Husen, E. D. I., (2013). Screening of rhizobacteria for plant growth promotion and their tolerance to drought stress. *Microbiol. Indones., 7*(3), 2.

Qiao, Q., Wang, F., Zhang, J., Chen, Y., Zhang, C., Liu, G., Zhang, H., Ma, C., & Zhang, J., (2017). The variation in the rhizosphere microbiome of cotton with soil type, genotype, and developmental stage. *Sci. Rep., 7*(1), 1–10.

Qurashi, A. W., & Sabri, A. N., (2011). Osmoadaptation and plant growth promotion by salt tolerant bacteria under salt stress. *African J. Microbiol. Res., 5*(21), 3546–3554.

Qurashi, A. W., & Sabri, A. N., (2012). Biofilm formation in moderately halophilic bacteria is influenced by varying salinity levels. *J. Basic Microbiol., 52*(5), 566–572.

Reddy, P. P., (2013). Recent advances in crop protection. *Recent Adv Crop Prot.*, 1–268. https://doi.org/10.1007/978-81-322-0723-8.

Riley, M. A., & Wertz, J. E., (2002). Bacteriocins: Evolution, ecology, and application. *Annu Rev Microbiol., 56*(1), 117–137.

Ryu, C. M., Farag, M. A., Hu, C. H., Reddy, M. S., Kloepper, J. W., & Paré, P. W., (2004). Bacterial volatiles induce systemic resistance in *Arabidopsis*. *Plant Physiol., 134*(3), 1017–1026.

Saeed, S., Ullah, A., Ullah, S., Noor, J., Ali, B., Khan, M. N., Hashem, M., et al., (2022). Validating the impact of water potential and temperature on seed germination of wheat (*Triticum aestivum* L.) via hydrothermal time model. *Life, 12*(7), 983.

Sandhya, V., Ali, S. Z., Grover, M., Reddy, G., & Venkateswarlu, B., (2010). Effect of plant growth promoting *Pseudomonas* spp. on compatible solutes, antioxidant status and plant growth of maize under drought stress. *Plant Growth Regul., 62*(1), 21–30.

Sandhya, V. Z. A. S., SK. Z, A., Grover, M., Reddy, G., & Venkateswarlu, B. S. S. S. (2009). Alleviation of drought stress effects in sunflower seedlings by the exopolysaccharides producing Pseudomonas putida strain GAP-P45. *Biology and Fertility of Soils, 46,* 17–26. https://doi.org/10.1007/s00374-009-0401-z.

Santoyo, G., Urtis-Flores, C. A., Loeza-Lara, P. D., Orozco-Mosqueda, M. D. C., & Glick, B. R., (2021). Rhizosphere colonization determinants by plant growth-promoting rhizobacteria (PGPR). *Biology (Basel), 10*(6), 475.

Savary, S., Willocquet, L., Pethybridge, S. J., Esker, P., McRoberts, N., & Nelson, A., (2019). The global burden of pathogens and pests on major food crops. *Nat. Ecol. Evol. [Internet], 3*(3), 430–439. https://doi.org/10.1038/s41559-018-0793-y.

Schippers, B., Bakker, A. W., & Bakker, P., (1987). Interactions of deleterious and beneficial rhizosphere microorganisms and the effect of cropping practices. *Annu. Rev. Phytopathol., 25*(1), 339–358.

Schuhegger, R., Ihring, A., Gantner, S., Bahnweg, G., Knappe, C., Vogg, G., Hutzler, P., et al., (2006). Induction of systemic resistance in tomato by N-acyl-L-homoserine lactone-producing rhizosphere bacteria. *Plant Cell Environ., 29*(5), 909–918.

Shagol, C. C., Subramanian, P., Krishnamoorthy, R., Kim, K., Lee, Y., Kwak, C., Sundaram, S., et al., (2014). ACC deaminase producing arsenic tolerant bacterial effect on mitigation of stress ethylene emission in maize grown in an arsenic polluted soil. *Korean J. Soil Sci. Fertil., 47*(3), 213–216.

Shanker, A., & Shanker, C. (Eds.). (2016). Abiotic and biotic stress in plants: recent advances and future perspectives. DOI: 10.5772/60477.

Sharma, C. K., Vishnoi, V. K., Dubey, R. C., & Maheshwari, D. K., (2018). A twin rhizospheric bacterial consortium induces systemic resistance to a phytopathogen *Macrophomina phaseolina* in mung bean. *Rhizosphere, 5*, 71–75.

Sharma, P., Khanna, V., & Kumari, P., (2013). Efficacy of aminocyclopropane-1-carboxylic acid (ACC)-deaminase-producing rhizobacteria in ameliorating water stress in chickpea under axenic conditions. *African J. Microbiol. Res., 7*(50), 5749–5757.

Singh, P. P., Shin, Y. C., Park, C. S., & Chung, Y. R., (1999). Biological control of *Fusarium* wilt of cucumber by chitinolytic bacteria. *Phytopathology, 89*(1), 92–99.

Singh, R. P., & Jha, P. N., (2016). A halotolerant bacterium *Bacillus licheniformis* HSW-16 augments induced systemic tolerance to salt stress in wheat plant (*Triticum aestivum*). *Front. Plant Sci., 7,* 1890.

Sivapriya, S. L., & Priya, P. R., (2017). Selection of hyper exopolysaccharide producing and cyst forming *Azotobacter* isolates for better survival under stress conditions. *Int. J. Curr. Microbiol. App. Sci., 6*(6), 2310–2320.

Sofy, A. R., Sofy, M. R., Hmed, A. A., & El-Dougdoug, N. K., (2019). Potential effect of plant growth-promoting rhizobacteria (PGPR) on enhancing protection against viral diseases. In: *F. Crop Sustain Manag by PGPR* (pp. 411–445). [place unknown]: Springer.

Sorty, A. M., Meena, K. K., Choudhary, K., Bitla, U. M., Minhas, P. S., & Krishnani, K. K., (2016). Effect of plant growth promoting bacteria associated with halophytic weed (*Psoralea corylifolia* L) on germination and seedling growth of wheat under saline conditions. *Appl. Biochem. Biotechnol., 180*(5), 872–882.

Subramanian, P., Mageswari, A., Kim, K., Lee, Y., & Sa, T., (2015). Psychrotolerant endophytic *Pseudomonas* sp. strains OB155 and OS261 induced chilling resistance in tomato plants (*Solanum lycopersicum* Mill.) by activation of their antioxidant capacity. *Mol. Plant-Microbe. Interact, 28*(10), 1073–1081.

Timmusk, S., Abd El-Daim, I. A., Copolovici, L., Tanilas, T., Kännaste, A., Behers, L., Nevo, E., et al., (2014). Drought-tolerance of wheat improved by rhizosphere bacteria from harsh environments: Enhanced biomass production and reduced emissions of stress volatiles. *PLoS One, 9*(5), e96086.

Upadhyay, S. K., Singh, J. S., Saxena, A. K., & Singh, D. P., (2012). Impact of PGPR inoculation on growth and antioxidant status of wheat under saline conditions. *Plant Biol., 14*(4), 605–611.

Uroz, S., Chhabra, S. R., Camara, M., Williams, P., Oger, P., & Dessaux, Y., (2005). N-Acylhomoserine lactone quorum-sensing molecules are modified and degraded by *Rhodococcus erythropolis* W2 by both amidolytic and novel oxidoreductase activities. *Microbiology, 151*(10), 3313–3322.

Uroz, S., Oger, P., Chhabra, S. R., Cámara, M., Williams, P., & Dessaux, Y., (2007). N-acyl homoserine lactones are degraded via an amidolytic activity in *Comamonas* sp. strain D1. *Arch. Microbiol., 187*(3), 249–256.

Verma, S. K., Kingsley, K., Bergen, M., English, C., Elmore, M., Kharwar, R. N., & White, J. F., (2018). Bacterial endophytes from rice cut grass (*Leersia oryzoides* L.) increase growth, promote root gravitropic response, stimulate root hair formation, and protect rice seedlings from disease. *Plant Soil, 422*(1), 223–238.

Voisard, C., Keel, C., Haas, D., & Dèfago, G., (1989). Cyanide production by *Pseudomonas fluorescens* helps suppress black root rot of tobacco under gnotobiotic conditions. *EMBO J., 8*(2), 351–358.

Wahab, A., Abdi, G., Saleem, M. H., Ali, B., Ullah, S., Shah, W., Mumtaz, S., et al., (2022). Alleviate the adverse effects of drought stress: A comprehensive review. *Plants, 11*, 1620.

Wang, W. Z., Morohoshi, T., Ikenoya, M., Someya, N., & Ikeda, T., (2010). AiiM, a novel class of N-acylhomoserine lactonase from the leaf-associated *bacterium Microbacterium testaceum*. *Appl. Environ. Microbiol., 76*(8), 2524–2530.

Wang, X., Mavrodi, D. V., Ke, L., Mavrodi, O. V., Yang, M., Thomashow, L. S., Zheng, N., et al., (2015). Biocontrol and plant growth-promoting activity of rhizobacteria from Chinese fields with contaminated soils. *Microb. Biotechnol., 8*(3), 404–418.

Weller, D. M., (2007). Pseudomonas biocontrol agents of soilborne pathogens: Looking back over 30 years. *Phytopathology, 97*(2), 250–256.

Weller, D. M., Mavrodi, D. V., Van, P. J. A., Pieterse, C. M. J., Van, L. L. C., & Bakker, P. A. H. M., (2012). Induced systemic resistance in *Arabidopsis thaliana* against *Pseudomonas syringae* pv. tomato by 2, 4-diacetylphloroglucinol-producing *Pseudomonas fluorescens. Phytopathology, 102*(4), 403–412.

Xun, F., Xie, B., Liu, S., & Guo, C., (2015). Effect of plant growth-promoting bacteria (PGPR) and arbuscular mycorrhizal fungi (AMF) inoculation on oats in saline-alkali soil contaminated by petroleum to enhance phytoremediation. *Environ. Sci. Pollut. Res., 22*(1), 598–608.

Yu, X., Ai, C., Xin, L., & Zhou, G., (2011). The siderophore-producing bacterium, *Bacillus subtilis* CAS15, has a biocontrol effect on *Fusarium* wilt and promotes the growth of pepper. *Eur. J. Soil Biol. [Internet]. 47*(2), 138–145. https://doi.org/https://doi.org/10.1016/j.ejsobi.2010.11.001.

Zaheer, A., Mirza, B. S., Mclean, J. E., Yasmin, S., Shah, T. M., Malik, K. A., & Mirza, M. S., (2016). Association of plant growth-promoting *Serratia* spp. with the root nodules of chickpea. *Res. Microbiol., 167*(6), 510–520.

Zaidi, S., Usmani, S., Singh, B. R., & Musarrat, J., (2006). Significance of *Bacillus subtilis* strain SJ-101 as a bioinoculant for concurrent plant growth promotion and nickel accumulation in *Brassica juncea. Chemosphere, 64*(6), 991–997.

Zainab, N., Khan, A. A., Azeem, M. A., Ali, B., Wang, T., Shi, F., Alghanem, S. M., et al., (2021). PGPR-mediated plant growth attributes and metal extraction ability of *Sesbania sesban* L. in industrially contaminated soils. *Agronomy, 11*(9), 1820.

Zhang, H., Kim, M. S., Sun, Y., Dowd, S. E., Shi, H., & Paré, P. W., (2008). Soil bacteria confer plant salt tolerance by tissue-specific regulation of the sodium transporter HKT1. *Mol. Plant-Microbe Interact., 21*(6), 737–744.

Zhang, H., Zhu, J., Gong, Z., & Zhu, J. K., (2022). Abiotic stress responses in plants. *Nat. Rev. Genet., 23*(2), 104–119.

Zheng, W., Zeng, S., Bais, H., LaManna, J. M., Hussey, D. S., Jacobson, D. L., & Jin, Y., (2018). Plant growth-promoting rhizobacteria (PGPR) reduce evaporation and increase soil water retention. *Water Resour. Res., 54*(5), 3673–3687.

CHAPTER 10

Adaptation Strategies for Salinity Stress in Maize: An Overview on Exogenous Chemical and Phyto-Microbiome Approaches

AYMAN EL SABAGH,[1,2] MOHAMMAD SOHIDUL ISLAM,[3] MUHAMMAD AAMIR IQBAL,[4] HUMAIRA YASMIN,[5] MOHSIN ZAFAR,[5] TAYYAB SHAFIQ,[5] KARTHIKA RAJENDRAN,[6] RUBAB SHABBIR,[7] TALHA JAVED,[7] MD. TAHJIB-UL-ARIF,[8] ARPNA KUMARI,[9] FATIH ÇIĞ,[2] SIPAN SOYSAL,[2] JAVEED LONE,[2] M. AMMAR RAZA,[2] and MURAT ERMAN[10]

[1]*Department of Agronomy, Faculty of Agriculture, Kafrelsheikh University, Kafrelsheikh, Egypt*

[2]*Department of Field Crops, Faculty of Agriculture, Siirt University, Siirt, Turkey*

[3]*Department of Agronomy, Hajee Mohammad Danesh Science and Technology University, Dinajpur, Bangladesh*

[4]*Department of Agronomy, Faculty of Agriculture, University of Poonch Rawalakot, Rawalakot, AJK, PakistanTechnology University, Dinajpur, Bangladesh*

[5]*COMSATS University Islamabad (CUI), Islamabad, Pakistan*

[6]*VIT School of Agricultural Innovations and Advanced Learning (VAIAL), Vellore Institute of Technology (VIT), Vellore, Tamil Nadu, India*

[7]*College of Agriculture, Fujian Agriculture and Forestry University, Fuzhou, China*

Advances in Plant Microbiome Research for Climate Resilient Agriculture: Toward Sustainable Farming. Ashwani Kumar (Ed.)

[6]*VIT School of Agricultural Innovations and Advanced Learning (VAIAL), Vellore Institute of Technology (VIT), Vellore, Tamil Nadu, India*

[7]*College of Agriculture, Fujian Agriculture and Forestry University, Fuzhou, China*

[8]*Plant Biology and Biofunctional Chemistry Lab, Department of Biochemistry and Molecular Biology, Bangladesh Agricultural University, Mymensingh, Bangladesh*

[9]*Academy of Biology and Biotechnology, Southern Federal University, Rostov-on-Don, Russia*

[10]*Department of Field Crops, Faculty of Agriculture, Bursa Uludağ University, Bursa, Turkey.*

ABSTRACT

Global warming is the leading cause of soil salinity, which affects the productivity of various crops, including maize. Soil salinity disrupts various physiological and biochemical processes in maize, including photosynthesis, water relations, respiration, stomatal regulation, and biosynthesis of enzymes. Improved performance of maize in saline soils may result from knowledge of the resistance to salt stress and an overview of the management options. Several approaches, alone or in combination have aided in the reduction of the negative effects of salinity by increasing salt tolerance in maize. Exogenous compounds, such as osmoprotectants (OP), polyamines (PA), nitric oxide (NO), and phytohormones, are thought to be the most promising and effective against environmental stresses, and they play critical roles in increasing maize resistance to salt stress. Furthermore, the use of microbes, which have been shown to be capable of reducing abiotic stresses, is a novel and promising method, and the plant microbiome has the ability to promote development and protect the host via a variety of molecular pathways. This chapter provides an overview of the exogenous application of several compounds and phyto-microbiomes that mitigate the negative effects of salt stress on maize.

10.1 INTRODUCTION

Climate is one of the most influential variables on essential activities and physiochemical properties in soil. Even though the weather has been drastically and continuously changing around the world, the effects on soil-related geological processes, modern agriculture, and human life are dynamic and dramatic. The average global temperature is expected to rise by 2 to 3°C by the end of this century (Abeer et al., 2015). Alarmingly, the harmful effects of changing climate scenarios on soil are predicted to be unidirectional, resulting in changes that vary in duration, occurrence (permanent or periodic), and consequence (favorable or unfavorable) (Várallyay, 2010). It has been discovered that atmospheric carbon dioxide enrichment, rising temperatures, more frequent and severe droughts, unequal and irregular precipitation due to altered rainfall patterns, and atmospheric nitrogen deposition all have a significant impact on soil functioning. Furthermore, there is growing evidence that crop yields in semi-arid and dry regions of the world may fall dramatically as a result of soil salinization and calcification (Zahra et al., 2018). Plants are subjected to a variety of stresses throughout their life cycle, most notably abiotic stresses such as salinity, heat, drought, and heavy metal toxicity. Crop plants respond differently to abiotic stimuli depending on their species, botanical traits, severity, and length (Sabagh el al., 2019). Among abiotic stresses, soil salinity has emerged as a major environmental stress and a major threat to the sustainability of modern agricultural systems. Numerous factors, such as soil fertility, soil condition, humidity, ambient temperature, light, and irrigation, interact, and tend to mitigate the negative effects of salinity. To determine a plant's tolerance to salinity, its ability to survive and thrive must generally be measured.

The saline environment alters several important morphological characteristics and physiological processes in crop plants, reducing productivity (Quintero et al., 2007). Salts such as sodium ions (Na^+), for example, can cause severe ionic toxicity because Na^+ ions are difficult to sequester in cell vacuoles (Fortmeier & Schubert, 1995). Salinity causes damage to plant cells under moderate stress, inhibiting growth. However, as salts accumulate in the adult leaves, photosynthesis ceases, and the plants begin to wilt. Furthermore, excessive salt can cause macronutrient and micronutrient imbalances. A high salinity environment causes a significant reduction in water intake, which disrupts the osmotic gradient between plants and soil. As a result, osmotic potentials drop, and plants are unable to absorb water (Munns et al., 2006). Thus, salinity stress causes physiological dryness and an increase in ionic toxicity. As a result, plant growth and development slow or stop, leaves show signs of wilting, necrosis, and chlorosis, and in extreme cases, the plant dies (Uddin et al., 2013).

After wheat and rice, maize (*Zea mays* L.) is the third most important and strategically pertinent crop (Maqsood et al., 2020). Its grain (8–10%) is used to manufacture a variety of human-edible products, including starch, glucose, corn oil, and a plant-based feed source (Khaliq et al., 2019). Furthermore, the fermentation process gets converted maize starch to biofuel (as ethanol). Maize is a naturally polymorphic, cross-pollinated plant (Iqbal et al., 2017). This crop is considered moderately salt-sensitive in general. However, salinity disrupts a number of physiological processes and growth properties in maize (Hasegawa et al., 2003; Rios-Gonzalez et al., 2002). Hormone regulation, respiration, transpiration, photosynthesis, stomatal functions, and water relations are all important physiological activities that prevent salinity from interfering with crop yield (Uddin et al., 2013). Furthermore, in a saline environment, phytotoxic ions will increase, usually causes oxidative stress that results in an increase in the production of reactive oxygen species (ROS) and ionic toxicity in the cytosol, reducing crop development and yield (Hichem et al., 2009).

Exogenous compounds appear to function as plant priming agents, boosting maize defense mechanisms and resilience to salt stress. Exogenous compounds such as OP, polyamines (PA), nitric oxide (NO), and phytohormones alleviate salt stress in maize. Furthermore, various agriculturally significant microbiomes have been extensively used for crop growth enhancement and stress control, particularly in maize. Plant-associated beneficial microbiomes have demonstrated that various methods can mitigate abiotic stressors (Vejan et al., 2016; Mapelli et al., 2013). Environmental challenges are posing an increasing threat to crop plant viability, necessitating nonconventional solutions such as the use of microbiomes (Schlaeppi & Bulgarelli, 2015). Future development of sustainable agriculture may involve enhancing beneficial microbial traits for plants, the environment, or both. The collection and use of microorganisms can be a crucial tool and important determinant in preserving plant health and productivity under various biotic and abiotic conditions (Berg et al., 2016; Marasco et al., 2016; Dastogeer et al., 2020). The discovery, characterization, and application of beneficial microbiomes that promote salinity stress tolerance in plants would contribute to the global agricultural sustainability of future generations (Nadeem et al., 2014; Jorquera et al., 2012). This chapter provided an overview of the exogenous chemical application techniques and use of microbiomes that enable maize to mitigate the detrimental effects of salt stress.

10.2 EXOGENOUS APPLICATION OF OSMOPROTECTANTS (OP)

Many studies have reported that numerous plant OP that can mitigate the negative effects of soil salinity on cereal crops. The OP, like glycine betaine (GB), is widely synthesized broadly in plants and plays an important role in conferring stress tolerance to crop plants (Tahjib-Ul-Arif et al., 2021; Sun et al., 2010; Ashraf & Foolad, 2007; Mäkelä et al., 1996). When environmental conditions are unfavorable, they tend to concentrate in higher plant chloroplasts and stabilize the quaternary structures of several complex proteins, such as PSII, which stops the breakdown of proteins (Murata et al., 1992). Furthermore, they protect cell membranes from higher Na^+ and Cl^- concentrations. In this way, OP helps plants cope with abiotic stresses like salinity, by preventing protein restructuring and membrane stabilization. However, in many economically important crops such as maize, OP naturally accumulates in much lower concentrations which is insufficient to mitigate the severe effects of salt stress (Subbarao et al., 2001).

Numerous factors influence the effectiveness of OP foliar application, including plant species, crop growth stage, GB dose, and number of foliar sprays. The factors have a greater impact on low osmoprotectant synthesizing crops like maize (Ashraf & Foolad, 2007). The accumulation of GB in maize plants in a saline environment remained 10-fold lower than salt-tolerant plants (Allard et al., 1998). However, maize cultivars that biosynthesize more OP were less affected by salinity than genotypes that synthesized no or a negligible amount of OP (Saneoka et al., 1995). It is widely assumed that exogenous application of OP modulates photosynthetic activity and antioxidant capacity, allowing plants to cope with salt stress. Under different salinity levels, maize genotypes were found to have reduced photosynthesis. At the same time, foliar-applied GB induced photosynthesis by increasing the CO_2 assimilation rate, which was closely related to crop salt tolerance. Foliar-applied GB increased endogenous GB levels by many orders of magnitude, which improved maize photosynthetic capacity by stomatal activity in a saline environment. Furthermore, OP like GB increased stomatal conductance, which helped to improve the net rate of photosynthesis, and plants were eventually able to grow in saline conditions (Rajasekaran et al., 1997, 1998). It has been reported that maize cultivars subjected to GB foliar application had higher photosynthetic rates due to stomatal conductance regulation by applied OP (Saneoka et al., 1995; Rajasekaran et al., 1997, 1998; Brugnoli & Björkman, 1992). Under saline conditions, leaf Na^+ concentrations increase significantly, while K^+, Ca^{2+}, and P concentrations decrease significantly in maize plants. The foliar spray of OP with K and P reduced the negative effects of salinity on maize by increasing K^+, Ca^{2+}, and P levels while decreasing photosynthetic pigment concentration and increasing the Na: K ratio (Kaya et al., 2013).

Typically, the underlying mechanisms of OP in maize increase photosynthetic capacity via stomatal regulation, accumulate endogenous OP, protect plasma membrane integrity, and protect transporter proteins as well as photosynthetic machinery. Another way OP confers salt tolerance on crop plants is by protecting the PSII protein complex by stabilizing O_2 in evolving centers (Murata et al., 1992). Furthermore, they maintain and promote enzyme activity (e.g., RUBISCO and phosphoenolpyruvate carboxylase) under salt stress, reducing salinity effects and increasing crops yield by 10–23% (Mäkelä et al., 1996).

Salinity has become a challenging task for maize production, as a result of climate change, particularly in arid and semi-arid regions of the world. In plants, foliar spraying of OP in conjunction with compatible osmolytes (proline) reduce the negative effects of salinity by improving water conservation while regulating stomatal conductance (Agboma et al., 1997). Furthermore, Farooq et al. (2009) concluded that foliar application of GB (100 mg L^{-1}) at the five-leaf stage remained effective in increasing crop tolerance to abiotic stresses (Farooq et al., 2009). However, significant research gaps exist regarding optimal doses of the various OP, the most appropriate time and growth stage of maize for foliar application, and the number of sprays required to effectively mitigate the adverse effects of salinity at varying levels. Furthermore, OP co-application with growth regulators (auxins, etc.) and minerals (N, P, K^+) requires extensive research to develop a powerful agronomic management strategy for salinity under changing climate scenarios.

10.3 EXOGENOUS APPLICATION OF POLYAMINES (PA)

Abiotic stresses, such as salinity, result in the biosynthesis of ROS such as hydroxyl radical ($HO^{\bullet}$), superoxide radical ($O_2^{\bullet-}$), and H_2O_2 (Schopfer et al., 2001). Thus, ROS plays an important role in many plants developmental processes under adverse environmental conditions (Demidchik & Maathuis, 2007). ROS, for example, activates the elongation zone in maize leaves, resulting in rapid leaf extension (Rodríguez et al., 2002). However, salinity causes an abrupt decrease in ROS biosynthesis, resulting in a severe decrease in leaf elongation and photosynthesis rate (Rodríguez et al., 2004).

The PA are small organic polycations that are naturally synthesized in various eukaryotic and prokaryotic cells and influence cell growth and the overall development of species (Steiner et al., 2007). The most abundant PAs in plant cells are spermidine (Spd), putrescine (Put), and spermine (Spm). Although PA has been found in both symplastic and apoplastic compartments, their primary biosynthesis site is in symplastic subcellular localizations (Tiburcio et al., 1997). Following biosynthesis and accumulation, PA begins to move from the plant cell membranes to the apoplast via an unknown mechanism. PA is catabolized in apoplasts by amine oxidases and other enzymes associated with apoplast compartments (Cona et al., 2006).

Exogenous polyamine application PA has recently revealed concrete evidence of their critical role in maize and other plant developmental processes (Walden et al., 1997). They remained effective in establishing tolerance in plants to salinity and other abiotic stresses (Kusano et al., 2007). Salt stress was found to increase the levels of apoplastic Spm and Spd, particularly in the elongation zone of the maize leaf blade. Polyamine oxidase activity was also found to be 20-fold higher than copper-containing amine oxidase. The estimation of H_2O_2, O_2^-, and HO· synthesis in the presence or absence of the polyamine oxidase inhibitors 1,19-bis-(ethylamine)-5,10,15 triazanonadecane and 1,8-diamino-octane indicates that in a saline environment, maize plants undergo oxidation of free apoplastic PA by polyamine oxidase, which served as the main source of ROS in the elongation zone of leaf blades. At the same time, this effect was attributed to enhance increased substrate availability (Rodríguez et al., 2009).

In terms of PA exogenous application, biotechnological approaches for PA-synthesizing enzymes overexpression could be developed to boost biosynthesis ROS biosynthesis within maize plants under abiotic stresses (Wi et al., 2006) as well as producing antisense and mutant generations with higher salt tolerance (Yamaguchi et al., 2006). It has been proposed that transgenic plants with overexpression of apoplastic polyamine oxidase extract H_2O_2 from PA catabolism (Moschou et al., 2008). More research is needed, however, to investigate the role of PA in boosting ROS biosynthesis and imparting salinity tolerance potential in maize crops under varying environmental conditions.

10.4 EXOGENOUS APPLICATION OF NITRIC OXIDE (NO)

Nitric oxide (NO) is involved in the regulation of many plant processors, particularly as a signaling molecule in response to biotic and abiotic stresses (Crawford & Guo, 2005). Exogenous NO application protects plants from stress-induced oxidative stress (Shi et al., 2005). Exogenous NO was used to regulate ROS accumulation, activate the antioxidant enzyme system, and control ROS damage under salinity stress (Xu et al., 2011). After applying NO under salinity, it was found that maize plants had less lipid peroxidation in their leaves (Klein et al., 2018). Salt-induced retarded activity of two O_2 generating enzymes, xanthine oxidase and NADPH oxidase resulting in reduced O_2 generation (Yun et al., 2011; Tewari et al., 2009). Superoxide production and lipid peroxidation were detected in maize leaves exposed to salinity. Exogenous NO application protects maize leaves from salinity-induced oxidative damage by activating superoxide dismutase (SOD) (Klein et al., 2018).

10.5 APPLICATION OF ARBUSCULAR MYCORRHIZAL FUNGI (AMF)

Arbuscular mycorrhizal fungi (AMF) are naturally occurring beneficial soil organisms capable of increasing crop productivity under a variety of environmental stresses, including salinity (Tahjib-Ul-Arif et al., 2018; Barea et al., 2013; Fa-Yuan, & Zhang, 2001). The symbiotic AMF-root association was found in nearly 80% of terrestrial plants (Smith & Read, 2008). AMF is important in decomposing soil litter and aggregating soil by binding soil organic matter, clay, and mineral ions and interacting with other soil microorganisms (Singh, 2016), forming fungal hyphae. AMF improved plant growth by increasing salt tolerance (Jahromi et al., 2008), improving water and mineral uptake, and enhancing ion homeostasis in saline environments (Giri et al., 2007; Asghari et al., 2005; Ruiz-Lozano & Azcón, 1995). Improved relative water content and water potential have been reported in various crop species, following AMF inoculation under salinity (Colla et al., 2008).

The effect of AMF inoculum on photosynthesis and water status under salt stress was investigated in a number of different crop species. According to reports, AMF colonization may increase the water potential of maize plants, the relative water content of zucchini leaves, and the chlorophyll concentration in the leaves of several plant species, including *Sesbania aegyptiaca*, *Sesbania grandiflora*, and *Lotus glaber* (Sannazzaro et al., 2006; Giri & Mukerji, 2004; Sheng et al., 2008). Moreover, AMF colonization is actively involved in soil mineral availability and solubility, increasing mineral uptake, particularly phosphorus, in maize plants in saline environments (Kazemi et al., 2019) by increasing phosphorus availability through activating the population of soil phosphorus solubilizing bacteria (Siciliano et al., 2014).

In saline soils, AMF can regulate Na^+ and K^+ translocation in roots keeping the shoot Na^+/K^+ ratio low (Farooq et al., 2015). Furthermore, AMF regulated Na^+ distribution, and ion compartmentalization in rice roots was observed in saline environments (Porcel et al., 2016). Furthermore, AMF induced photosynthesis-related events such as high net carbon gain, increased leaf area, enhanced photochemical reactions, and preventing oxidative damage in maize grown in saline soils (Labidi et al., 2011; Omoto et al., 2012).

10.6 APPLICATION OF PLANT GROWTH PROMOTING RHIZOBIA

Plant growth-promoting rhizobia (PGPR) are beneficial microorganisms that live in the rhizosphere and are particularly effective at reducing salinity in crop plants (Shahid et al., 2018). In numerous crop species, including rice (Kumar et al., 2017), sweet pepper (Wang et al., 2018), Bassia (Abeer et al., 2015), radish (Mohamed & Gomaa, 2012), *Arabidopsis* (Ortíz-Castro et al., 2017), and sorghum (Mahmood et al., 2017), PGPR species belonging to the genus *Bacillus* increased growth in response to salinity stress. PGPR enhanced water intake and water usage efficiency in saline conditions, hence mitigating physiological drought caused by osmosis (Kohler et al., 2009). It has been reported that PGPR inoculated maize grown in r saline soils produces more plant biomass (Abd et al., 2004). Due to a decrease in chlorophyll concentration and Rubisco activity, salinity may have an effect on photosynthesis (Nakano et al., 2006). However, *Bacillus* SR-2-1/1 species upregulated Rubisco activity, resulting in improved CO_2 fixation and enhanced photosynthetic rate. *Bacillus* SR-2-1/1 species maintain the K^+/Na^+ ratio and compartmentalize excess Na^+ in vacuoles by overexpressing the Na^+/H^+ tonoplast antiporter gene *NHX1*, activating other transporters such as *SOS1* to maintain Na^+ homeostasis and exclude other excessive ions in maize (Rafiq et al., 2020). Moreover, methods of ROS scavenging mediated by PGPR-have been identified in salinity sensitive cells. Upon inoculation with PGPR, *Arthrobacter nitroguajacolicus*, it activates catalase (CAT) and peroxides cycles (Safdarian et al., 2019). Thus, the use of PGPR inoculants could aid in mitigating the harm caused by salinity to maize.

10.7 EXOGENOUS APPLICATION OF PHYTOHORMONES

Hormones are the most important chemicals involved in increased plant growth and productivity under extreme or normal conditions. Plant hormones, such as auxin, ABA, ethylene, cytokinin, gibberellins, salicylic acid (SA), brassinosteroids (BR), and strigolactones help plants adapt to harsh environmental conditions. Because of their different actions, all of these hormones play distinct roles (Miransari, 2016). The effect is most noticeable during differentiation, expansion, and maturation (Kaya et al., 2009). Hormones are the primary regulators of plant growth and development; they are small, simple molecules with varying chemical compositions. They might be endogenous or exogenous.

ABA is an important phytohormone that helps plants cope with stress. This hormone has long been linked to a lack of water in the soil around the root. Because salinity stress causes osmotic stress and a lack of water, ABA production in shoots and roots would improve salinity tolerance. Other compounds, such as BR and SA, also play a role in plant abiotic stress response. Under conditions of high salinity, there was an increase in the endogenous level of SA, as well an increase in the activity of the enzyme responsible for biosynthesizing SA (Gupta & Huang, 2014). Salinity is the primary challenge to maize cultivation around the world. Understanding its responses to saline conditions allows for the development of strategies for high production. Previous research has shown that germination is more sensitive in saline environments.

Furthermore, externally applied hormones tend to improve the toxic effect of salt stress in maize by retaining turgor, improving nutrient consumption, and maintaining membrane and stability. Salt stress reduces total biomass, chlorophyll a and chlorophyll b levels, and electrolyte level leakage. Thus, the use of kinetin and IAA reversed the harmful effect of NaCl stress while having no effect on maize resistance to salinity. As a result, we can improve maize performance under salt conditions by using exogenous hormones and OP. The use of these substances allows for osmotic adaptation, nutrient uptake, and antioxidant defense (Farooq et al., 2015).

Under adverse conditions, the plant growth regulators play an important role in modulating physiological responses to adapt an unfavorable environment. In particular, cytokines control chlorophyll and delay leaf senescence, two crucial functions in plant growth and development (Prasad, 2022). 6-Benzyladenine, often known as 6-BA, belongs to the cytokine category of plant growth regulators (Rohwer & Erwin, 2015). Synthetic plant growth can be induced with 6-BA. 6-BA may boost cytokine levels in plants, inhibiting and free scavenge radicals, prevent pith tissue and leaf aging and prolong active photosynthetic life. Additionally, 6-BA may improve the plant resistance to environmental stresses such as drought, flooding, salt stress and low-temperature stress (Ren et al., 2016). However, their roles in reducing the negative effects of waterlogging on ear characteristics, grain filling, dry matter distribution, photosynthesis, and chlorophyll content have received little attention. In the study we present here, a field experiment was conducted to determine how 6-BA spraying affected summer maize yield in the sixth and third leaf stages. This study evaluated its potential to enhance morpho biochemical properties. The physiological properties of maize during a drought highlight the likelihood of using two exogenous growth regulators, SA, and methyl jasmonate (MeJA). Stress decelerated maize development (shoot and root lengths; dry and fresh weights), and the negative effects of MeJA and SA treatments were lessened by stress (Xiaotao et al., 2013). Drought stress can drastically reduce maize yield and interfere with photosynthetic, feedback, metabolic and developmental processes. In a recent study, the exogenous use of SA and MeJA, in comparison to untreated controls, has mitigated the detrimental effects of drought stress in maize (Javid et al., 2011). Applications of MeJA led to the maximum rise in endogenous ABA during drought stress. However, the detrimental effects of oxidative stress brought on drought have been less effectively mitigated by SA and MeJA combined half-dose procedures. The treatment increased osmolytes to the greatest extent possible, maintaining the water status in maize plants and enhancing antioxidant activity under drought stress. Exogenous treatment of SA and MeJA as seed and foliar pre-treat combinations may, in some cases, modify osmolytes, endogenous hormones, and antioxidant enzymes, preventing the negative effects of osmotic and oxidative stress caused by drought (Yuan et al., 2018).

10.8 CONCLUSION

The changing climate scenario has increased soil salinity due to drought, which causes salts to move upward with water from deeper layers to soil horizons. Saline or salt-affected soils have an excess of soluble salts or exchangeable sodium in the root zone. Salt stress has emerged as one of the most serious threats to modern commercial-oriented maize production systems in arid and semi-arid regions, owing to low rainfall and high evapotranspiration caused by rising temperatures caused by global warming, combined with ineffective soil and water management practices. The prime toxic ion that interferes with potassium uptake and disrupts stomatal modulation, resulting in significant water losses and necrosis in maize in sodium. To enable crop plants to lessen the negative impacts of saline environment, considerable efforts must be made in light of the changing climate and growing saline areas around the world. The deleterious effects soil salinity on maize plants may be lessened by the exogenous application of antioxidants, PA, NO, PGPRs, and different hormones. However, there is a critical need for more study in the areas of source and dose optimization, application timing, agricultural plant growth stages, etc.

KEYWORDS

- **abiotic stress**
- **arbuscular mycorrhizal fungi**
- **mycorrhizal fungi**
- **nitric oxide**
- **osmoprotectants**
- **phytomicrobiome**
- **plant response**
- ***Zea mays* L.**

REFERENCES

Abd, M., Hamdia, E. S., Shaddad, M. A. K., & Doaa, M. M., (2004). Mechanisms of salt tolerance and interactive effects of *Azospirillum brasilense* inoculation on maize cultivars grown under salt stress conditions. *Plant Growth Regulation, 44*(2), 165–574. doi: 10.1023/B:GROW.0000049414.03099.9B.

Abeer, H., Abd-Allah, E. F., Alqarawi, A. A., et al., (2015). Impact of plant growth promoting *Bacillus subtilis* on growth and physiological parameters of *Bassia indica* (*Indian Bassia*) grown udder salt stress. *Pak J Bot., 47*(5), 1735–5741.

Agboma, P. C., Jones, M. G. K., Peltonen-Sainio, P., Rita, H., & Pehu, E., (1997). Exogenous glycinebetaine enhances grain yield of maize, sorghum and wheat grown under two supplementary watering regimes. *J. Agron. Crop. Sci., 178*(1), 29–37. doi: 10.1111/J.1439-037X.1997.TB00348.X.

Allard, F., Houde, M., Kröl, M., Ivanov, A., Huner, N. P. A., & Sarhan, F., (1998). Betaine improves freezing tolerance in wheat. *Plant Cell Physiol., 39*(11), 1194–4202. doi: 10.1093/OXFORDJOURNALS.PCP.A029320.

Asghari, H. R., Marschner, P., Smith, S. E., & Smith, F. A., (2005). Growth response of *Atriplex nummularia* to inoculation with arbuscular mycorrhizal fungi at different salinity levels. *Plant and Soil, 273*(1), 245–556. doi: 10.1007/S11104-004-7942-6.

Ashraf, M., & Foolad, M. R., (2007). Roles of glycine betaine and proline in improving plant abiotic stress resistance. *Environ. Exp Bot., 59*(2), 206–616. doi: 10.1016/J.ENVEXPBOT.2005.12.006.

Barea, J., Pozo, M., López-Ráez, J., et al., (2013). Arbuscular mycorrhizas and their significance in promoting soil-plant system sustainability against environmental stresses. *Beneficial Plant-Microbial Interactions, 2013*, 353–387. doi: 10.1201/B15251-16.

Berg, G., Rybakova, D., Grube, M., & Köberl, M., (2016). The plant microbiome explored: Implications for experimental botany. *J. Exp. Bot., 67*(4), 995–5002. doi: 10.1093/JXB/ERV466.

Brugnoli, E., & Björkman, O., (1992). Growth of cotton under continuous salinity stress: Influence on allocation pattern, stomatal and non-stomatal components of photosynthesis and dissipation of excess light energy. *Planta, 187*(3), 335–547. doi: 10.1007/BF00195657.

Colla, G., Rouphael, Y., Cardarelli, M., Tullio, M., Rivera, C. M., & Rea, E., (2008). Alleviation of salt stress by arbuscular mycorrhizal in zucchini plants grown at low and high phosphorus concentration. *Biol. Fertil. Soils, 44*(3), 501–109. doi: 10.1007/S00374-007-0232-8/TABLES/3.

Cona, A., Rea, G., Botta, M., Corelli, F., Federico, R., & Angelini, R., (2006). Flavin-containing polyamine oxidase is a hydrogen peroxide source in the oxidative response to the protein phosphatase inhibitor cantharidin in *Zea mays* L. *J. Exp. Bot., 57*(10), 2277–7289. doi: 10.1093/JXB/ERJ195.

Crawford, N. M., & Guo, F. Q., (2005). New insights into nitric oxide metabolism and regulatory functions. *Trends Plant Sci., 10*(4), 195–500. doi: 10.1016/J.TPLANTS.2005.02.008.

Dastogeer, K. M. G., Zahan, M. I., Tahjib-Ul-Arif, M., Akter, M. A., & Okazaki, S., (2020). Plant salinity tolerance conferred by arbuscular mycorrhizal fungi and associated mechanisms: A meta-analysis. *Front Plant Sci., 11*, 1927. doi: 10.3389/FPLS.2020.588550/BIBTEX.

Demidchik, V., & Maathuis, F. J. M., (2007). Physiological roles of nonselective cation channels in plants: From salt stress to signaling and development. *New Phytol., 175*(3), 387–704. doi: 10.1111/J.1469-9137.2007.02128.X.

Farooq, M., Basra, S. M. A., Wahid, A., Ahmad, N., & Saleem, B. A., (2009). Improving the drought tolerance in rice (*Oryza sativa* L.) by exogenous application of salicylic acid. *J. Agron. Crop Sci., 195*(4), 237–746. doi: 10.1111/J.1439-037X.2009.00365.X.

Farooq, M., Hussain, M., Wakeel, A., & Siddique, K. H. M., (2015). Salt stress in maize: Effects, resistance mechanisms, and management. A review. *Agron. Sustain. Dev., 35*(2), 461–181. doi: 10.1007/S13593-015-0287-0/TABLES/2.

Fa-Yuan, W., & Zhang, Y., (2001). A preliminary survey of arbuscular mycorrhizal fungi in saline alkaline soil of the yellow River Delta. *Biodiv. Sci., 9*(4), 389. doi: 10.17520/BIODS.2001057.

Fortmeier, R., & Schubert, S., (1995). Salt tolerance of maize (*Zea mays* L.): The role of sodium exclusion. *Plant Cell Environ., 18*(9), 1041–1047. doi: 10.1111/J.1365-5040.1995.TB00615.X.

Giri, B., & Mukerji, K. G., (2004). Mycorrhizal inoculant alleviates salt stress in *Sesbania aegyptiaca* and *Sesbania grandiflora* under field conditions: Evidence for reduced sodium and improved magnesium uptake. *Mycorrhiza., 14*(5), 307–712. doi: 10.1007/S00572-003-0274-1.

Giri, B., Kapoor, R., & Mukerji, K. G., (2007). Improved tolerance of *Acacia nilotica* to salt stress by arbuscular mycorrhiza, *Glomus fasciculatum* may be partly related to elevated K/Na ratios in root and shoot tissues. *Microb. Ecol., 54*(4), 753–360. doi: 10.1007/S00248-007-9239-9.

Gupta, B., & Huang, B., (2014). Mechanism of salinity tolerance in plants: Physiological, biochemical, and molecular characterization. *Int J Genomics, 2014*. doi: 10.1155/2014/701596.

Hasegawa, P. M., Bressan, R. A., Zhu, J. K., & Bohnert, H. J., (2003). *Plant Cellular and Molecular Responses to High Salinity, 51*, 463–399. http://dx.doi.org/101146/annurev.arplant511463. doi: 10.1146/ANNUREV.ARPLANT.51.1.463.

Hichem, H., Mounir, D., & Naceur, E. A., (2009). Differential responses of two maize (*Zea mays* L.) varieties to salt stress: Changes on polyphenols composition of foliage and oxidative damages. *Ind. Crops Prod., 30*(1), 144–451. doi: 10.1016/J.INDCROP.2009.03.003.

Iqbal, M. A., Aslam, Z., Maqsood, M., Ahmad, Z., Akbar, N., ... & Faisal, M. (2017). Boosting forage yield and quality of *maize* (Zea mays L.) with multi-species bacterial inoculation in Pakistan. Phyton, 86, 84. doi: 10.32604/PHYTON.2017.86.084.

Jahromi, F., Aroca, R., Porcel, R., & Ruiz-Lozano, J. M., (2008). Influence of salinity on the *in vitro* development of *Glomus intraradices* and on the *in vivo* physiological and molecular responses of mycorrhizal lettuce plants. *Microb Ecol., 55*(1), 45–53. doi: 10.1007/S00248-007-9249-7.

Javid, M. G., Sorooshzadeh, A., Sanavy, S. A. M. M., Allahdadi, I., & Moradi, F., (2011). Effects of the exogenous application of auxin and cytokinin on carbohydrate accumulation in grains of rice under salt stress. *Plant Growth Regul., 65*(2), 305–513. doi: 10.1007/S10725-011-9602-1/FIGURES/3.

Jorquera, M. A., Shaharoona, B., Nadeem, S. M., De La Luz, M. M., & Crowley, D. E., (2012). Plant growth-promoting rhizobacteria associated with ancient clones of creosote bush (*Larrea tridentata*). *Microb. Ecol., 64*(4), 1008–8017. doi: 10.1007/S00248-012-0071-5.

Kaya, C., Ashraf, M., Dikilitas, M., & Tuna, A. L., (2013). Alleviation of salt stress-induced adverse effects on maize plants by exogenous application of indoleacetic acid (IAA) and inorganic nutrients-A field trial. *Aust. J. Crop Sci., 7*(2), 249–954.

Kaya, C., Tuna, A. L., & Yokaş, I., (2009). *The Role of Plant Hormones in Plants Under Salinity Stress* (pp. 45–50). doi: 10.1007/978-1-4020-9065-3_5.

Kazemi, R., Ronaghi, A., Yasrebi, J., Ghasemi-Fasaei, R., & Zarei, M., (2019). Effect of shrimp waste–derived biochar and arbuscular mycorrhizal fungus on yield, antioxidant enzymes, and chemical composition of corn under salinity stress. *J. Soil Sci. Plant Nutr., 19*(4), 758–870. doi: 10.1007/S42729-019-00075-2.

Khaliq, A., Iqbal, M. A., Zafar, M., & Gulzar, A., (2019). Appraising economic dimension of maize production under coherent fertilization in Azad Kashmir, Pakistan. *Custos E Agronegocio on Line, 15*(2), 243–353. https://eurekamag.com/research/070/936/070936999.php (accessed on 21 October 2023).

Klein, A., Hüsselmann, L., Keyster, M., & Ludidi, N., (2018). Exogenous nitric oxide limits salt-induced oxidative damage in maize by altering superoxide dismutase activity. *South African J. Bot., 115*, 44–49. doi: 10.1016/J.SAJB.2017.12.010.

Kohler, J., Hernández, J. A., Caravaca, F., & Roldán, A., (2009). Induction of antioxidant enzymes is involved in the greater effectiveness of a PGPR versus AM fungi with respect to increasing the tolerance of lettuce to severe salt stress. *Environ. Exp. Bot., 65*(2, 3), 245–552. doi: 10.1016/J.ENVEXPBOT.2008.09.008.

Kumar, K., Amaresan, N., & Madhuri, K., (2017). Alleviation of the adverse effect of salinity stress by inoculation of plant growth promoting rhizobacteria isolated from hot humid tropical climate. *Ecol. Eng., 102*, 361–166. doi: 10.1016/J.ECOLENG.2017.02.023.

Kusano, T., Yamaguchi, K., Berberich, T., & Takahashi, Y., (2007). Advances in polyamine research in 2007. *J. Plant Res., 120*(3), 345–550. doi: 10.1007/S10265-007-0074-3.

Labidi, S., Calonne, M., Ben, J. F., et al., (2011). Calcareous impact on arbuscular mycorrhizal fungus development and on lipid peroxidation in monoxenic roots. *Phytochem., 72*(18), 2335–5341. doi: 10.1016/J.PHYTOCHEM.2011.08.016.

Mahmood, F., Shahid, M., Hussain, S., et al., (2017). Potential plant growth-promoting strain *Bacillus* sp. SR-2-1/1 decolorized azo dyes through NADH-ubiquinone: Oxidoreductase activity. *Bioresour. Technol., 235*, 176–684. doi: 10.1016/J.BIORTECH.2017.03.098.

Mäkelä, P., Mantila, J., Hinkkanen, R., Pehu, E., & Peltonen-Sainio, P., (1996). Effect of foliar applications of glycinebetaine on stress tolerance, growth, and yield of spring cereals and summer turnip rape in Finland. *J Agron. Crop Sci., 176*(4), 223–334. doi: 10.1111/J.1439-037X.1996.TB00467.X.

Mapelli, F., Marasco, R., Rolli, E., et al., (2013). Potential for plant growth promotion of rhizobacteria associated with *Salicornia* growing in Tunisian hypersaline soils. *Biomed Res. Int., 2013*. doi: 10.1155/2013/248078.

Maqsood, Q., Abbas, R. N., Iqbal, M. A., Serap, K. A., Iqbal, A., & Sabagh, A. E., (2020). Overviewing of weed management practices to reduce weed seed bank and to increase maize yield. *Planta Daninha., 38*, 1–10. doi: 10.1590/S0100-83582020380100075.

Marasco, R., Mapelli, F., Rolli, E., et al., (2016). *Salicornia strobilacea* (synonym of *Halocnemum strobilaceum*) grown under different tidal regimes selects rhizosphere bacteria capable of promoting plant growth. *Front Microbiol.*, *7*, 1286. doi: 10.3389/FMICB.2016.01286/BIBTEX.

Miransari, M., (2016). Soybeans, stress, and nutrients. *Environmental Stresses in Soybean Production: Soybean Production*, 2:273–398. doi: 10.1016/B978-0-12-801535-3.00012-7.

Mohamed, H. I., & Gomaa, E. Z., (2012). *Effect of Plant Growth Promoting Bacillus Subtilis and Pseudomonas Fluorescens on Growth and Pigment Composition of Radish Plants (Raphanus Sativus) Under NaCl Stress, 50*(2), 263–372. http://ps.ueb.cas.cz/doi/101007/s11099-012-0032-8.html. doi: 10.1007/S11099-012-0032-8.

Moschou, P. N., Paschalidis, K. A., Delis, I. D., et al., (2008). Spermidine exodus and oxidation in the apoplast induced by abiotic stress is responsible for H_2O_2 signatures that direct tolerance responses in tobacco. *Plant Cell, 20*(6), 1708. doi: 10.1105/TPC.108.059733.

Munns, R., James, R. A., & Läuchli, A., (2006). Approaches to increasing the salt tolerance of wheat and other cereals. *J. Exp. Bot., 57*(5), 1025–5043. doi: 10.1093/JXB/ERJ100.

Murata, N., Mohanty, P. S., Hayashi, H., & Papageorgiou, G. C., (1992). Glycinebetaine stabilizes the association of extrinsic proteins with the photosynthetic oxygen-evolving complex. *FEBS Lett., 296*(2), 187–789. doi: 10.1016/0014-5793(92)80376-R.

Nadeem, S. M., Ahmad, M., Zahir, Z. A., Javaid, A., & Ashraf, M., (2014). The role of mycorrhizae and plant growth promoting rhizobacteria (PGPR) in improving crop productivity under stressful environments. *Biotechnol. Adv., 32*(2), 429–948. doi: 10.1016/J.BIOTECHADV.2013.12.005.

Nakano, R., Ishida, H., Makino, A., & Mae, T., (2006). *In vivo* fragmentation of the large subunit of ribulose-1,5-bisphosphate carboxylase by reactive oxygen species in an intact leaf of cucumber under chilling-light conditions. *Plant Cell Physiol., 47*(2), 270–076. doi: 10.1093/PCP/PCI245.

Omoto, E., Taniguchi, M., & Miyake, H., (2012). Adaptation responses in C4 photosynthesis of maize under salinity. *J Plant Physiol., 169*(5), 469–977. doi: 10.1016/J.JPLPH.2011.11.009.

Ortíz-Castro, R., Valencia-Cantero, E., & López-Bucio, J., (2008). Plant growth promotion by *Bacillus megaterium* involves cytokinin signaling. *Plant Signal Behav., 3*(4), 263. doi: 10.4161/PSB.3.4.5204.

Photosynthesis Under Drought Stress. https://www.researchgate.net/publication/282569319_Photosynthesis_under_drought_stress (accessed on 21 October 2023).

Porcel, R., Aroca, R., Azcon, R., & Ruiz-Lozano, J. M., (2016). Regulation of cation transporter genes by the arbuscular mycorrhizal symbiosis in rice plants subjected to salinity suggests improved salt tolerance due to reduced Na(+) root-to-shoot distribution. *Mycorrhiza., 26*(7), 673–384. doi: 10.1007/S00572-016-0704-5.

Prasad, R. (2022). Cytokinin and its key role to enrich the plant nutrients and growth under adverse conditions-an update. *Frontiers in Genetics, 13*, 883924. doi: 10.3389/fgene.2022.883924.

Quintero, J. M., María, F. J., & Benlloch, M., (2007). Na+ accumulation in shoot is related to water transport in K+-starved sunflower plants but not in plants with a normal K+ status. *J Plant Physiol., 164*(1), 60–67. doi: 10.1016/J.JPLPH.2005.10.010.

Rafiq, K., Akram, M. S., Shahid, M., Qaisar, U., & Rashid, N., (2020). Enhancement of salt tolerance in maize (*Zea mays* L.) using locally isolated *Bacillus* sp. SR-2-1/1. *Biologia (Bratisl), 75*(9), 1425–5436. doi: 10.2478/S11756-020-00435-9/FIGURES/2.

Rajasekaran, L. R., Kriedemann, P. E., Aspinall, D., & Paleg, L. G., (1998). Physiological significance of proline and glycinebetaine: Maintaining photosynthesis during NaCl stress in wheat. *Photosynthetica, 34*(3), 357–766. doi: 10.1023/A:1006855816437.

Ren, B., Zhu, Y., Zhang, J., Dong, S., Liu, P., & Zhao, B., (2016). Effects of spraying exogenous hormone 6-benzyladenine (6-BA) after waterlogging on grain yield and growth of summer maize. *Field Crops Res.*, (188), 96–104. doi: 10.1016/J.FCR.2015.10.016.

Rios-Gonzalez, K., Erdei, L., & Lips, S. H., (2002). The activity of antioxidant enzymes in maize and sunflower seedlings as affected by salinity and different nitrogen sources. *Plant Sci., 162*(6), 923–330. doi: 10.1016/S0168-9452(02)00040-7.

Rodríguez, A. A., Córdoba, A. R., Ortega, L., & Taleisnik, E., (2004). Decreased reactive oxygen species concentration in the elongation zone contributes to the reduction in maize leaf growth under salinity. *J. Exp. Bot., 55*(401), 1383–3390. doi: 10.1093/JXB/ERH148.

Rodríguez, A. A., Grunberg, K. A., & Taleisnik, E. L., (2002). Reactive oxygen species in the elongation zone of maize leaves are necessary for leaf extension. *Plant Physiol., 129*(4), 1627–7632. doi: 10.1104/PP.001222.

Rodríguez, A. A., Maiale, S. J., Menéndez, A. B., & Ruiz, O. A., (2009). Polyamine oxidase activity contributes to sustain maize leaf elongation under saline stress. *J. Exp. Bot., 60*(15), 4249–9262. doi: 10.1093/JXB/ERP256.

Rohwer, C. L., & Erwin, J. E., (2015). Horticultural Applications of Jasmonates, *83*(3), 283–304. http://dx.doi.org/101080/14620316200811512381.

Ruiz-Lozano, J. M., & Azcón, R., (1995). Hyphal contribution to water uptake in mycorrhizal plants as affected by the fungal species and water status. *Physiol Plant., 95*(3), 472–278. doi: 10.1111/J.1399-9054.1995.TB00865.X.

Sabagh, A. E., Hossain, A., Barutçular, C., Islam, M. S., Ratnasekera, D., Kumar, N., ... & da Silva, J. A. T. (2019). Drought and salinity stress management for higher and sustainable canola ('Brassica napus' L.) production: A critical review. *Aust. J Crop Sci., 13*(1), 88-96. doi: 10.21475/ajcs.19.13.01.p1284.

Safdarian, M., Askari, H., Shariati, J. V., & Nematzadeh, G., (2019). Transcriptional responses of wheat roots inoculated with *Arthrobacter nitroguajacolicus* to salt stress. *Sci. Rep. 9*(1), 1–12. doi: 10.1038/s41598-018-38398-2.

Saneoka, H., Nagasaka, C., Hahn, D. T., et al., (1995). Salt tolerance of glycinebetaine-deficient and -containing maize lines. *Plant Physiol., 107*(2), 631. doi: 10.1104/PP.107.2.631.

Sannazzaro, A. I., Ruiz, O. A., Albertó, E. O., & Menéndez, A. B., (2006). Alleviation of salt stress in *Lotus glaber* by *Glomus intraradices*. *Plant Soil, 285*(1), 279–987. doi: 10.1007/S11104-006-9015-5.

Schlaeppi, K., & Bulgarelli, D., (2015). The plant microbiome at work. *Mol. Plant Microbe. Interact., 28*(3), 212–217. doi: 10.1094/MPMI-10-14-0334-FI.

Schopfer, P., Plachy, C., & Frahry, G., (2001). Release of reactive oxygen intermediates (superoxide radicals, hydrogen peroxide, and hydroxyl radicals) and peroxidase in germinating radish seeds controlled by light, gibberellin, and abscisic acid. *Plant Physiol., 125*(4), 1591–1602. doi: 10.1104/PP.125.4.1591.

Shahid, M., Akram, M. S., Khan, M. A., et al., (2018). A phytobeneficial strain *Planomicrobium* sp. MSSA-10 triggered oxidative stress responsive mechanisms and regulated the growth of pea plants under induced saline environment. *J. Appl. Microbiol., 124*(6), 1566–6579. doi: 10.1111/JAM.13732.

Sheng, M., Tang, M., Chen, H., Yang, B., Zhang, F., & Huang, Y., (2008). Influence of arbuscular mycorrhizae on photosynthesis and water status of maize plants under salt stress. *Mycorrhiza, 18*(6, 7), 287–796. doi: 10.1007/S00572-008-0180-7/TABLES/2.

Shi, S., Wang, G., Wang, Y., Zhang, L., & Zhang, L., (2005). Protective effect of nitric oxide against oxidative stress under ultraviolet-B radiation. *Nitric Oxide, 13*(1), 1-9. doi: 10.1016/J.NIOX.2005.04.006.

Siciliano, S. D., Palmer, A. S., Winsley, T., et al., (2014). Soil fertility is associated with fungal and bacterial richness, whereas pH is associated with community composition in polar soil microbial communities. *Soil Biol. Biochem., 78*, 10–20. doi: 10.1016/J.SOILBIO.2014.07.005.

Singh, K., (2016). Microbial and enzyme activities of saline and sodic soils. *Land Degrad Dev., 27*(3), 706–618. doi: 10.1002/LDR.2385.

Smith, S. E., & Read, D. J., (2008). Ericoid, orchid and mycoheterotrophic mycorrhizas. *Mycorrhizal Symbiosis* (pp. 387–707). Published online.

Steiner, N., Santa-Catarina, C., Silveira, V., Floh, E. I. S., & Guerra, M. P., (2007). Polyamine effects on growth and endogenous hormones levels in *Araucaria angustifolia* embryogenic cultures. *Plant Cell, Tissue, and Organ Culture (PCTOC), 1*(89), 55–62. doi: 10.1007/S11240-007-9216-5.

Subbarao, G. V., Wheeler, R. M., Levine, L. H., & Stutte, G. W., (2001). Glycine betaine accumulation, ionic and water relations of red-beet at contrasting levels of sodium supply. *J. Plant Physiol., 158*(6), 767–776. doi: 10.1078/0176-1617-00309.

Sun, Y. L., Hong, S. K., Sun, Y. L., & Hong, S. K., (2010). Exogenous proline mitigates the detrimental effects of saline and alkaline stresses in *Leymus chinensis* (Trin.). *J. Plant Biotechnol., 37*(4), 529–938. doi: 10.5010/JPB.2010.37.4.529.

Tahjib-Ul-Arif, M., Ghosh, A., Chamely, S. G., Haque, M. R., & Rahman, M. M., (2018). Arbuscular mycorrhizal fungi inoculation with organic matter and phosphorus supplementation enhance nutrient contents of *Amaranthus tricolor* L. and *Basella alba* L. by improving nutrients uptake. *The J. Soc. Trop. Plant Res.* doi: 10.22271/tpr.2018.v5.i3.046.

Tahjib-Ul-Arif, M., Shorna, S. I., Sadik, P. M. A., et al., (2021). Role of organic solutes on mineral stress tolerance. *Organic Solutes, Oxidative Stress, and Antioxidant Enzymes Under Abiotic Stressors* (pp. 83–104). doi: 10.1201/9781003022879-4.

Tewari, R. K., Kumar, P., Kim, S., Hahn, E. J., & Paek, K. Y., (2009). Nitric oxide retards xanthine oxidase-mediated superoxide anion generation in *Phalaenopsis* flower: An implication of NO in the senescence and oxidative stress regulation. *Plant Cell Rep., 28*(2), 267–779. doi: 10.1007/S00299-008-0632-8/FIGURES/9.

Tiburcio, A. F., Altabella, T., Borrell, A., & Masgrau, C., (1997). Polyamine metabolism and its regulation. *Physiol. Plant., 100*(3), 664–474. doi: 10.1111/J.1399-9054.1997.TB03073.X.

Uddin, M. N., Hanstein, S., Leubner, R., & Schubert, S., (2013). Leaf cell-wall components as influenced in the first phase of salt stress in three maize (*Zea mays* L.) hybrids differing in salt resistance. *J. Agron. Crop Sci., 199*(6), 405–515. doi: 10.1111/JAC.12031.

Várallyay, G., (2010). The impact of climate change on soils and on their water management. *Agronomy Research, 8*(Special II), 385–596.

Vejan, P., Abdullah, R., Khadiran, T., Ismail, S., & Nasrulhaq, B. A., (2016). Role of plant growth promoting rhizobacteria in agricultural sustainability: A review. *Mole., 21*(5). doi: 10.3390/MOLECULES21050573.

Walden, R., Cordeiro, A., & Tiburcio, A. F., (1997). Polyamines: Small molecules triggering pathways in plant growth and development. *Plant Physiol., 113*(4), 1009. doi: 10.1104/PP.113.4.1009.

Wang, W., Wu, Z., He, Y., Huang, Y., Li, X., & Ye, B. C., (2018). Plant growth promotion and alleviation of salinity stress in *Capsicum annuum* L. by *Bacillus* isolated from saline soil in Xinjiang. *Ecotoxicol. Environ. Saf., 164*, 520–029. doi: 10.1016/J.ECOENV.2018.08.070.

Wi, S. J., Kim, W. T., & Park, K. Y., (2006). Overexpression of carnation S-adenosylmethionine decarboxylase gene generates a broad-spectrum tolerance to abiotic stresses in transgenic tobacco plants. *Plant Cell Rep., 25*(10), 1111–1121. doi: 10.1007/S00299-006-0160-3.

Xiaotao, D., Yuping, J., Hong, W., et al., (2013). Effects of cytokinin on photosynthetic gas exchange, chlorophyll fluorescence parameters, antioxidative system and carbohydrate accumulation in cucumber (*Cucumis sativus* L.) under low light. *Acta Physiol. Plant, 35*(5), 1427–7438. doi: 10.1007/S11738-012-1182-9.

Xu, J., Yin, H., Yang, L., Xie, Z., & Liu, X., (2011). Differential salt tolerance in seedlings derived from dimorphic seeds of *Atriplex centralasiatica*: From physiology to molecular analysis. *Planta., 233*(5), 859–971. doi: 10.1007/S00425-010-1347-Y.

Yamaguchi, K., Takahashi, Y., Berberich, T., et al., (2006). The polyamine spermine protects against high salt stress in *Arabidopsis thaliana*. *FEBS Lett., 580*(30), 6783–3788. doi: 10.1016/J.FEBSLET.2006.10.078.

Yuan, C., Feng, S., Wang, J., et al., (2018). Effects of irrigation water salinity on soil salt content distribution, soil physical properties and water use efficiency of maize for seed production in arid Northwest China. *Int. J. Agril. Biol. Eng., 11*(3), 137–745. doi: 10.25165/IJABE.V11I3.3146.

Yun, B. W., Feechan, A., Yin, M., et al., (2011). S-nitrosylation of NADPH oxidase regulates cell death in plant immunity. *Nature, 478*(7368), 264–468. doi: 10.1038/NATURE10427.

Zahra, N., Mahmood, S., & Raza, Z. A., (2018). *Salinity Stress on Various Physiological and Biochemical Attributes of Two Distinct Maize (Zea Mays L.) Genotypes*, *41*(11), 1368–8380. https://doi.org/101080/0190416720181452939. doi: 10.1080/01904167.2018.1452939.

CHAPTER 11

The Beneficial Plant Microbial Association for Sustainable Agriculture

EEMAN FATIMA,[1] MUHAMMAD AMMAR JAVED,[2] AQSA HAFEEZ,[3] and BABER ALI[3]

[1]*Department of Botany, Government College University, Lahore, Pakistan*

[2]*Institute of Industrial Biotechnology, Government College University, Lahore, Pakistan*

[3]*Department of Plant Sciences, Quaid-i-Azam University, Islamabad, Pakistan*

ABSTRACT

Microorganisms are ubiquitous and can interact with each other to colonize plants and form different modes of association. These microorganisms could colonize both the rhizosphere and phyllosphere to aggravate plant growth in various ways and enhance their resistance capabilities. These microorganisms, also called plant growth-promoting microorganisms (PGPMs), lessen the toxic effect of chemical fertilizers and pesticides and reduce their demand in agriculture. Microorganisms form such associations in which they provide nutrition, mitigate various abiotic stresses as well, and provide resistance against variegate pathogens directly or indirectly. Some species form direct symbiotic associations with plants while others form associative symbiosis or non-symbiotic beneficial interactions like siderophores production, growth hormones and vitamin synthesis, water and mineral translocation, protection against nematodes and pathogens, solubilization, nutrient acquisition, and resistance against abiotic and biotic stresses. This chapter deals with the types of associations that develop between microorganisms and plants and

Advances in Plant Microbiome Research for Climate Resilient Agriculture: Toward Sustainable Farming. Ashwani Kumar (Ed.)

how these interactions benefit each other. Moreover, the classification of PGP-rhizobacteria will also be highlighted.

11.1 INTRODUCTION

The interaction between plants and microorganisms is a complex, continuous, and dynamic process that is millions of years old (Ali et al., 2022). The microbial association is an interaction between host and non-host species forming an ecological unit (Ali et al., 2022). The abiotic factors such as cold, temperature (Saeed et al., 2022), drought (Bibi et al., 2022; Dola et al., 2022; Farooq et al., 2022; Wahab et al., 2022), heavy metals (Ma, Ali et al., 2022; Ma, Saleem, Ali et al., 2022; Ma, Saleem, Yasin et al., 2022; Nawaz et al., 2022; Naz et al., 2022; Saleem et al., 2022), salt concentration (Ali et al., 2022; Khan et al., 2022; Hussain et al., 2023), pesticides (Anwar et al., 2023), biotic factors (Asad et al., n.d.; Metayi et al., 2022) drastically affect crop yield and productivity. Plants are frequently invaded by helpful and harmful microorganisms, mostly bacteria and fungus, in both natural and farmed settings (Wille et al., 2019). Fungus roots have been identified as a unique adaption of plant roots that exist in around 85% of all plant species on Earth. According to recent estimations, 263 families and 3,617 plant species may be associated with mycorrhiza. As a result, the mycorrhizal state is thought to be the most common symbiotic condition on earth. The advantageous interactions can be described as several direct or indirect mechanisms, such as nutrient transfer, which is carried out by mycorrhizal fungi and rhizobia, which associate with roots and provide plants with mineral nutrients (Adnan et al., 2022) and fixed nitrogen, respectively, direct stimulation of growth through phytohormones, antagonistic activity against pathogenic microorganisms (Solanki et al., 2022), and stress reduction (Mehmood et al., 2021; Zainab et al., 2021). Conversely, unfavorable interactions injure plants because the encroaching microorganisms may be saprophytic and result in necrotrophy in the colonizing plants (Al-Zaban et al., 2022). Understanding how plants and microorganisms interact is therefore essential to understanding both the beneficial and harmful effects that microbes have on plants (Dolatabadian, 2020; Afridi, Ali et al., 2022; Afridi, Javed et al., 2022).

11.2 MYCORRHIZAS

The root of the plant and the fungus form a mycorrhizal connection. The juxtaposition of a root and a spore accelerates spore germination in reaction to root exudates, and fungal hyphae and spores in soil can function as inocula for roots. The signaling mechanisms involved in host/fungal recognition are currently being revealed by research. When a plant and a fungus recognize their compatibility, the fungi will associate with the root to form the morphological structures unique to that particular symbiosis if the environmental conditions are right (Dighton, 2003).

The main types of mycorrhizas are as follows:

- Endomycorrhiza; and
- Ectomycorrhiza.

11.2.1 ENDOMYCORRHIZA

Endomycorrhiza, as its name indicates, is a collective term for all mycorrhizal connections in which the fungus is mostly internal to the root structure and penetrates the cortical cell walls of the host. The arbuscular and ericoid mycorrhizae are the two main types; however additional organisms, like the dark septate endophytes, may also fall under this heading. There is hyphal penetration into the host cells in each of these mycorrhizal types.

11.2.2 ARBUSCULAR MYCORRHIZAE

Arbuscular mycorrhizae are a type of symbiotic association where hydrostatic pressure and enzyme activity are used to break through the epidermal cells and fungal hyphae leave behind an appressorium, or hyphal swelling, on the root surface where pressure builds up. The hyphae then break through the plasma membrane of the cortical cell walls, pushing it aside, and branch into distinctive arbuscules, which resemble trees, to increase the surface area of contact between the fungus and the contents of the host cell (Bever et al., 2005). The symbiotic relationship between the fungal and plant components is made easier by the huge surface area. The Arum and Paris forms of arbuscule formation are the two most common varieties. The Paris type forms its arbuscules from vast hyphal coils inside the host cells, whereas the Arum type generates multibranched hyphal structures that resemble tree-like arbuscules. This group of mycorrhizae is made up of a significant number of vascular plant species, such as grasses, herbs, and trees, especially tropical tree species, and a relatively small number of fungus species (about 150) from the phylum Glomeromycota (Koide & Mosse, 2004).

11.2.3 ERICOID MYCORRHIZAE

In the Ericaceae, Epacridaceae, and Empetraceae, a small group of fungi known as ericoid mycorrhizae is connected to a small variety of plant species. The first fungus species identified as an ericaceous endosymbiont was *Hymenoscyphus (Pezizella) ericae*. However, more recently, it has been shown that several additional fungal species, including Oidiodendron, Myxotrichium, and Gymnascella, develop mycorrhizal connections with ericoid plants. The fungal hyphae infiltrate cortical cells, often of extremely thin roots, in a way similar to arbuscular mycorrhizae, although they do not form arbuscules but rather hyphal coils. The vascular bundle and one layer of cortical/epidermal cells make up the fine roots of ericoid plants.

11.2.4 ECTOMYYCORHIZA

In ectomycorrhiza, both the fungal hyphal sheath and mantle that surrounds the exterior of the root and the extraradical hyphae and hyphal structures extend into the surrounding soil to make up the symbiotic association. As their name implies, ectomycorrhizae have a substantial amount of the biomass of their fungi partners outside the root. The complexity of the sheath can range from a loose weave of hyphae to a dense and organized multilayer of cells that resembles plant parenchyma tissue; this is known as pseudoparenchymatous. The juxtaposition of fungal and plant symbionts in this area, known as the Hartig net, is where nutrient and carbon exchange take place. By inducing altered amounts of cytokinins, which lead to greater branching, the colonization of root tips by ectomycorrhizal fungi often inhibits the growth of root hairs.

A variety of Basidiomycotina, Ascomycotina, and certain Zygomycotina, of which 5,500 species have been identified as myco-symbionts, may contribute to the formation of the ectomycorrhizal connection. Numerous common forest and woodland mushrooms and truffles, such as Russula, Hebeloma, Cortinaria, Lactarius, Laccaria, Amanita, and Lycoperdon, have fungal companions (e.g., Tuber). In all biomes, these fungi are associated with a small number of tree species. Most coniferous trees, larch, birch, beech, oak, and eucalyptus are among the genera of trees.

11.2.5 ORCHID MYCORRHIZA

Orchid mycorrhizae might be seen as the perfect example of how plants and fungi are interdependent. There are almost 17,000 different types of orchids, and they all rely on their basidiomycetes fungus partners to obtain nutrition. Within the cortical cells of the host plant, the fungus forms highly coiled arbuscules (peletons). These fungus coils have a limited lifespan and leave behind pectin and cellulose in the host cell when they die. New hyphae that can reach these carbohydrate and food sources may subsequently "invade" these cells.

11.2.6 ARBUTOID

Arbutus and Arctostaphylos, two distinct members of the Ericaceae, including several Pyrolaceae genera are involved in this particular mycorrhizal connection. These arbutoid relationships are created by a variety of ectomycorrhizal fungal species and feature a thin fungal sheath covering the outside of the root, a paraepidermal Hartig net created by fungal penetration between the epidermis, and an outer layer of cortical cells into which the hyphae invade to create hyphal coils (intracellular hyphal complexes). Arbutoid plants and nearby tree species may share nutrients and photosynthesizes, according to some research.

11.2.7 MONOTROPOID

These mycorrhizal kinds are only found in the Ericaceae family's Monotropoideae, which are achlorophyllous plants that dwell on the forest floor and have mostly lost their ability to photosynthesize. Originally believed to be parasitic on some types of forest trees, it is now discovered that they share mycorrhizal symbionts with their nearby trees. These plants use mycorrhizal bridges between their root systems to receive carbohydrates from nearby trees. These connections have been linked to the fungi Tricholoma, Russula, and Rhizopogon. Similar to the ectomycorrhizal state on trees, these fungi create fungal sheaths and Hartig nets on their monotropoid hosts.

11.2.8 DARK SEPTATE ENDOPHYTES

Many plant species from practically all plant families, especially those living in chilly, nutritionally deficient conditions, have been shown to have these root endophytes. However, roots have allowed for the identification of individuals belonging to the genera Chloridium, Leptodontidium, Phialocephala, and Phialophora. These fungi enter the root through root hairs or cortical cells, and after establishing runner hyphae between cortical cells, the host cell is invaded by dense, multibranched structures known as microsclerotia. Hyaline hyphae frequently deposit melanin in their cell walls at first, giving these fungi their name for their dark color.

11.3 MYCORRHIZAL FUNCTION

The mycorrhizal function is traditionally thought to promote the host plant's intake of inorganic nutrients from the soil (Fahad et al., 2022). However, there is evidence that mycorrhizal connection provides various host advantages, so this may be oversimplifying their function. The main functions of the symbiotic association are as follows.

Mycorrhizal association between fungus and plant's root is of great concern. Mycorrhizal fungi associated with plant roots increase the absorption of nutrients, particularly phosphorus, and thus enhance the growth of crop plants and trees (Ahmad et al., 2022). The fungi colonize the root system of a host plant, providing increased water and nutrient absorption capabilities while the plant provides the fungus with carbohydrates formed from photosynthesis.

11.4 PLANT GROWTH-PROMOTING RHIZOBACTERIA (PGPR)

Plant growth-promoting rhizobacteria (PGPR) are rhizosphere microorganisms that can improve plant growth through many different mechanisms, including phosphate solubilization, siderophore production, biological nitrogen fixation, rhizosphere engineering, production of 1-Aminocyclopropane-1-carboxylate deaminase (ACC), interference with quorum sensing (QS) signaling and inhibition of biofilm formation, As PGPR offers a desirable alternative to the use of chemical fertilizers, pesticides, and other additives, its potential in agriculture is rapidly growing (Amna et al., 2021; Ali, Wang, Saleem, Azeem et al., 2022; Ali, Wang, Saleem, Hafeez et al., 2022).

These rhizosphere bacteria are expected to generate growth-promoting chemicals in significant amounts, which might have an indirect impact on the overall morphology of the plants. Recent advancements in our knowledge of the variety of PGPR in the rhizosphere, as well as their capacity for colonization and mode of action, should make it easier to apply them as a dependable part of the management of sustainable agricultural systems (Saini et al., 2022; Salam et al., 2022).

11.5 CLASSIFICATION

PGPRs may be divided into two categories: extracellular (ePGPR) and intracellular PGPR (iPGPR) (Martínez-Viveros et al., 2010). The iPGPRs are often found inside the specialized nodular structures of root cells, but the ePGPRs can be found in the rhizosphere, on the rhizoplane, or in the gaps between the cells of the root cortex. Agrobacterium, Arthrobacter, Azotobacter, Azospirillum, Bacillus, Burkholderia, Caulobacter, Chromobacterium, Erwinia, Flavobacterium, Micrococcus, Pseudomonas, and Serratia are among the bacterial genera that belong to the ePGPR (Gray & Smith, 2005).

The endophytes and Frankia species, both of which may symbiotically fix atmospheric N_2 with higher plants, are included in the iPGPR (Verma et al., 2010). A vast variety of soil bacterial species, known as endophytes, such as Allorhizobium, Azorhizobium, Bradyrhizobium, Mesorhizobium, and Rhizobium of the family Rhizobiaceae often penetrate the root systems of agricultural plants to create nodules, which either directly or indirectly encourage growth. Gram-positive rods, cocci, and pleomorphic rhizobacteria make up a smaller percentage of this type of rhizobacteria, which are primarily Gram-negative and rod-shaped. Examples include *Allorhizobium undicola* (de Lajudie et al., 1998), *Azorhizobium caulinodans* (Dreyfus et al., 1988), *Bradyrhizobium japonicum* (Guerinot & Chelm, 1984), *Mesorhizobium chacoense* (Velazquez et al., 2001), *Mesorhizobium pluriforium* (Scholla & Elkan, 1984).

Researchers have looked at the possibility that PGPR help tomatoes and peppers withstand water stress (Mayak et al., 2004). Fluorescent pseudomonads and Bacillus species have been discovered to colonize host roots extremely effectively and produce growth compounds that increase the output of critical crops (Khalid et al., 2004). According to Kilian et al. (2000), a *Bacillus subtilis* strain called FZB24's diverse mechanisms of action against phytopathogens points to the bacterium's contribution to plant life.

By colonizing the rhizosphere, soil rhizobacterial populations can have a positive impact on a variety of plants, including wheat, potato, maize, grasses, pea, and cucumber, according to Çakmakçi et al. (2006). Over a broad range of root zone temperatures (RZTs), PGPR applications boosted soya bean (*Glycine max* (L. Merr.) nodulation and nitrogen fixation (Zhang et al., 1996). Thus, it has been demonstrated that the inoculation of PGPRs can enhance agricultural plants' nodulation, absorption of nitrogen, growth, and yield response. Another significant field of study in the past 10 years has been on using microorganisms as co-cultures in biotization (Sekar & Kandavel, 2010).

An increase in the ability of plants to withstand biotic and abiotic stress (Faryal et al., 2022; Saleem et al., 2022) is brought about by biotization, which is a metabolic reaction of *in vitro* produced plant material to microbial inoculants that results in developmental and physiological alterations of the generated propagules. In this case, plantlets and PGPR are often co-cultured to increase biomass and secondary metabolite production. For instance, plantlets of *Origanum vulgare L.* generated higher phenolics and chlorophyll when co-cultured with *Pseudomonas* spp. than the non-bacterized control (Nowak, 1998).

11.6 ENDOPHYTIC BACTERIAL ASSOCIATIONS

Numerous studies have described the possible benefits of plant-associated bacteria as factors promoting plant development and preserving the health of the soil and plants. The endophytes in plants are overlooked due to several reasons, including their modest size, variety, and culturable nature. Because of this, plant biologists view all plants as one organism. As a result, the latter's vital contributions were occasionally disregarded or ignored. Most plant species inevitably harbor endophytic bacteria, which have been isolated from a range of plants. Endophytic bacteria are thought to be a largely untapped prospective resource for the discovery of isolates with new antibiotic compounds and PGP properties among plant-associated microorganisms (Lodewyckx et al., 2002; Rosenblueth & Martínez-Romero, 2006).

Because of their close relationship with plants, bacterial endophytes present a special opportunity for their prospective use in biological control and plant protection. Although the biocontrol capabilities of microorganisms involving the synthesis of allelochemicals have primarily been studied in the context of free-living rhizobacteria, endophytic bacteria are also capable of producing metabolites that have antagonistic activity toward plant pathogens (Lodewyckx et al., 2002). An agricultural ecosystem's total production and the defense of plants against soil-borne diseases may be significantly impacted by certain endophytic bacterial isolates (Hallmann et al., 1997; Sturz et al.,

2000). Through the de novo synthesis of structural compounds and the inhibition of fungal penetration, the establishment of stable microbial endoplant communities may result in disease resistance (Benhamou & Nicole, 1999), the induction and expression of general molecular-based plant immunity, or the straightforward exclusion of other organisms (phytopathogens and colonists) through niche competition (Sturz et al., 2000). Competition for an ecological niche or a substrate, the generation of inhibitory allelochemicals, and the induction of systemic resistance (ISR) in host plants to a wide range of diseases and/or abiotic stressors are the generally acknowledged processes of biocontrol mediated by both species.

However, the host specificity, population dynamics, pattern of host colonization, ability to migrate within host tissues, and capacity to generate systemic resistance, all play a role in how successful endophytes are as biological control agents (Backman et al., 1997). For instance, Chen et al. (1995) discovered five endophytic species that could significantly lessen the vascular wilt in cotton brought on by *Fusarium oxysporum f.* sp. *vasinfectum: Aureobacterium saperdae, Bacillus pumilus, Phyllobacterium rubiacearum, Pseudomonas putida*, and *Burkholderia solanacearum*. Divergent hosts might be colonized, as shown by *Pseudomonas* sp. strain PsJN, an onion endophyte, which suppressed *Botrytis cinerea Pers*. and encouraged vine development in colonized grapevines (Barka et al., 2002).

It can be concluded that to preserve and improve global food production in future climatic scenarios while maintaining environmental health, it is crucial to increase the activities of microorganisms that boost plant nutrition, manage diseases, and help plants to cope with a range of abiotic challenges. Although a wide variety of helpful microbes have been discovered, their effective usage in field settings has not yet been completely realized.

KEYWORDS

- **dark septate endophytes**
- **endophytic bacterial associations**
- **ericoid mycorrhizae**
- **monotropoid**
- **mycorrhizal function**
- **plant growth-promoting rhizobacteria**

REFERENCES

Adnan, M., Fahad, S., Saleem, M. H., Ali, B., Mussart, M., Ullah, R., Arif, M., et al., (2022). Comparative efficacy of phosphorus supplements with phosphate solubilizing bacteria for optimizing wheat yield in calcareous soils. *Sci. Rep., 12*(1), 1–17.

Afridi, M. S., Ali, S., Salam, A., César, T. W., Hafeez, A., Ali, B., AlTami, M. S., et al., (2022). Plant microbiome engineering: Hopes or hypes. *Biology (Basel). 11*(12), 1782.

Afridi, M. S., Javed, M. A., Ali, S., De Medeiros, F. H. V., Ali, B., Salam, A., Sumaira, Marc, R. A., et al., (2022). New opportunities in plant microbiome engineering for increasing agricultural sustainability under stressful conditions. *Front. Plant Sci., 13*, 1–22. https://doi.org/10.3389/fpls.2022.899464.

Ahmad, M., Ishaq, M., Shah, W. A., Adnan, M., Fahad, S., Saleem, M. H., Khan, F. U., Mussarat, M., Khan, S., Ali, B., et al., (2022). Managing phosphorus availability from organic and inorganic sources for optimum wheat production in calcareous soils. *Sustain., 14*(13), 7669. https://doi.org/10.3390/su14137669.

Ali, B., Hafeez, A., Ahmad, S., Javed, M. A., Afridi, M. S., Dawoud, T. M., Almaary, K. S., et al., (2022). *Bacillus thuringiensis* PM25 ameliorates oxidative damage of salinity stress in maize via regulating growth, leaf pigments, antioxidant defense system, and stress responsive gene expression. *Front. Plant Sci.*, 13.

Ali, B., Hafeez, A., Javed, M. A., Afridi, M. S., Abbasi, H. A., Qayyum, A., Batool, T., Ullah, A., Marc, R. A., Jaouni, S. K. Al., et al., (2022). Role of endophytic bacteria in salinity stress amelioration by physiological and molecular mechanisms of defense: A comprehensive review. *South African J. Bot., 151*, 33–46. https://doi.org/10.1016/j.sajb.2022.09.036.

Ali, B., Wang, X., Saleem, M. H., Azeem, M. A., Afridi, M. S., Nadeem, M., Ghazal, M., et al., (2022). Bacillus mycoides PM35 reinforces photosynthetic efficiency, antioxidant defense, expression of stress-responsive genes, and ameliorates the effects of salinity stress in maize. *Life, 12*(2), 219.

Ali, B., Wang, X., Saleem, M. H., Hafeez, A., Afridi, M. S., Khan, S., Ullah, I., et al., (2022). PGPR-mediated salt tolerance in maize by modulating plant physiology, antioxidant defense, compatible solutes accumulation and bio-surfactant producing genes. *Plants, 11*(3), 345.

Ali, S., Ullah, S., Khan, M. N., Khan, W. M., Razak, S. A., Wahab, S., Hafeez, A., et al., (2022). The effects of osmosis and thermo-priming on salinity stress tolerance in *Vigna radiata* L. *Sustain., 14*(19), 12924. https://doi.org/10.3390/su141912924.

Al-Zaban, M. I., Alhag, S. K., Dablool, A. S., Ahmed, A. E., Alghamdi, S., Ali, B., Al-Saeed, F. A., et al., (2022). Manufactured nano-objects confer viral protection against cucurbit chlorotic yellows virus (CCYV) infecting *Nicotiana benthamiana. Microorganisms [Internet], 10*(9), 10. https://doi.org/10.3390/microorganisms10091837.

Amna, Ali, B., Azeem, M. A., Qayyum, A., Mustafa, G., Ahmad, M. A., Javed, M. T., & Chaudhary, H. J., (2021). Bio-fabricated silver nanoparticles: A sustainable approach for augmentation of plant growth and pathogen control. In: *Sustain. Agric. Rev.* (Vol. 53, pp. 345–371). [place unknown]: Springer.

Anwar, T., Qureshi, H., Perveen, N., Mahmood, S., Zulqurnain, H. M., Mumtaz, S., Nawaz, H., Khan, S. A., Hafeez, A., Imran, T. M., et al., (2023). Herbicidal effectiveness of wild poisonous plant *Rhazya stricta* using different media by the sandwich method. *Pakistan J. Bot., 55*(2), 2. https://doi.org/10.30848/pjb2023-2(10).

Asad, M., Khan, R. R., Aljuboory, A. B., Rashid, M. H. U., Kumar, U., Haq, I. U., ... & Alharbi, K. (2023). Toxic and Antifeedant Effects of Different Pesticidal Plant Extracts against Beet. Phyton, 92(4).

Backman, P. A., Wilson, M., & Murphy, J. F. (2018). Bacteria for biological control of plant diseases. In *Environmentally Safe Approaches to Crop Disease Control* (pp. 94-110). CRC Press.

Barka, E. A., Gognies, S., Nowak, J., Audran, J. C., & Belarbi, A., (2002). Inhibitory effect of endophyte bacteria on *Botrytis cinerea* and its influence to promote the grapevine growth. *Biol. Control, 24*(2), 135–142.

Benhamou, N., & Nicole, M., (1999). Cell biology of plant immunization against microbial infection: The potential of induced resistance in controlling plant diseases. *Plant Physiol. Biochem., 37*(10), 703–719.

Bever, J. D., Schultz, P. A., Dighton, J., White, J., & Oudemans, P., (2005). Mechanisms of arbuscular mycorrhizal mediation of plant-plant interactions. *Mycol. Ser., 23*, 443.

Bibi, S., Ullah, S., Hafeez, A., Khan, M. N., Javed, M. A., Ali, B., Din, I. U., et al., (2022). Exogenous Ca/Mg quotient reduces the inhibitory effects of PEG induced osmotic stress on *Avena sativa* L. *Brazilian J. Biol.,* 84.

Çakmakçi, R., Dönmez, F., Aydın, A., & Şahin, F., (2006). Growth promotion of plants by plant growth-promoting rhizobacteria under greenhouse and two different field soil conditions. *Soil Biol. Biochem., 38*(6), 1482–1487.

Chen, C., Bauske, E. M., Musson, G., Rodriguezkabana, R., & Kloepper, J. W., (1995). Biological control of *Fusarium* wilt on cotton by use of endophytic bacteria. *Biol. Control., 5*(1), 83–91.

De Lajudie, P., Willems, A., Nick, G., Moreira, F., Molouba, F., Hoste, B., Torck, U., et al., (1998). Characterization of tropical tree rhizobia and description of *Mesorhizobium plurifarium* sp. Nov. *Int. J. Syst. Evol. Microbiol., 48*(2), 369–382.

Dighton, J., (2003). *Fungi in Ecosystem Processes*. New York: M.

Dola, D. B., Mannan, M. A., Sarker, U., Mamun, M. A. A., Islam, T., Ercisli, S., Saleem, M. H., et al., (2022). Nano-iron oxide accelerates growth, yield, and quality of *Glycine max* seed in water deficits. *Front. Plant Sci., 13*, 992535.

Dolatabadian, A., (2020). Plant–microbe interaction. *Biology (Basel), 10*(1), 15.

Dreyfus, B., Garcia, J. L., & Gillis, M., (1988). Characterization of *Azorhizobium caulinodans* gen. Nov., sp. Nov., a stem-nodulating nitrogen-fixing bacterium isolated from *Sesbania rostrata. Int. J. Syst. Bacteriol., 38*(1), 89–98.

Fahad, S., Chavan, S. B., Chichaghare, A. R., Uthappa, A. R., Kumar, M., Kakade, V., Pradhan, A., Jinger, D., Rawale, G., Yadav, D. K., et al., (2022). Agroforestry systems for soil health improvement and maintenance. *Sustainability, 14*(22). https://doi.org/10.3390/su142214877.

Farooq, T. H., Rafay, M., Basit, H., Shakoor, A., Shabbir, R., Riaz, M. U., ... & Jaremko, M. (2022). Morpho-physiological growth performance and phytoremediation capabilities of selected xerophyte grass species toward Cr and Pb stress. *Frontiers in Plant Science, 13*, 997120.

Faryal, S., Ullah, R., Khan, M. N., Ali, B., Hafeez, A., Jaremko, M., & Qureshi, K. A., (2022). Thiourea-capped nanoapatites amplify osmotic stress tolerance in *Zea mays* L. by conserving photosynthetic pigments, osmolytes biosynthesis and antioxidant biosystems. *Molecules, 27*(18), 5744. https://doi.org/10.3390/molecules27185744.

Gray, E. J., & Smith, D. L., (2005). Intracellular and extracellular PGPR: Commonalities and distinctions in the plant–bacterium signaling processes. *Soil Biol. Biochem., 37*(3), 395–412.

Guerinot, M. L., & Chelm, B. K., (1984). Isolation and expression of the *Bradyrhizobium japonicum* adenylate cyclase gene (cya) in *Escherichia coli. J. Bacteriol., 159*(3), 1068–1071.

Hallmann, J., Quadt-Hallmann, A., Mahaffee, W. F., & Kloepper, J. W., (1997). Bacterial endophytes in agricultural crops. *Can. J. Microbiol., 43*(10), 895–914.

Hussain, S. S., Rasheed, M., Hamzah, S. M., Ahmed, Z. I., Hafeez, A., Jilani, G., Alamri, S., et al., (2023). Salt tolerance in maize with melatonin priming to achieve sustainability in yield on salt affected soils. *Pakistan J. Bot., 55*(1). https://doi.org/10.30848/pjb2023-1(27).

Khalid, A., Arshad, M., & Zahir, Z. A., (2004). Screening plant growth-promoting rhizobacteria for improving growth and yield of wheat. *J. Appl. Microbiol., 96*(3), 473–480.

Khan, M. A., Adnan, M., Basir, A., Fahad, S., Hafeez, A., Saleem, M. H., Ahmad, M., et al., (2022). Impact of tillage and potassium levels and sources on growth, yield, and yield attributes of wheat. *Pak. J. Bot., 55*(10.30848).

Kilian, M., Steiner, U., Krebs, B., Junge, H., Schmiedeknecht, G., & Hain, R. (2000). FZB24® Bacillus subtilis–mode of action of a microbial agent enhancing plant vitality. *Pflanzenschutz-Nachrichten Bayer, 1*, 1.

Koide, R. T., & Mosse, B., (2004). A history of research on arbuscular mycorrhiza. *Mycorrhiza, 14*, 145–163.

Lodewyckx, C., Vangronsveld, J., Porteous, F., Moore, E. R. B., Taghavi, S., Mezgeay, M., & Der Lelie, D. V., (2002). Endophytic bacteria and their potential applications. *CRC Crit. Rev. Plant Sci., 21*(6), 583–606.

Ma, J., Ali, S., Saleem, M. H., Mumtaz, S., Yasin, G., Ali, B., Al-Ghamdi, A. A., Elshikh, M. S., Vodnar, D. C., Marc, R. A., et al., (2022). Short-term responses of spinach (*Spinacia oleracea* L.) to the individual and combinatorial effects of nitrogen, phosphorus and potassium and silicon in the soil contaminated by boron. *Front Plant Sci., 13*. https://doi.org/10.3389/fpls.2022.983156.

Ma, J., Saleem, M. H., Ali, B., Rasheed, R., Ashraf, M. A., Aziz, H., Ercisli, S., Riaz, S., Elsharkawy, M. M., Hussain, I., et al., (2022). Impact of foliar application of syringic acid on tomato (*Solanum lycopersicum* L.) under heavy metal stress-insights into nutrient uptake, redox homeostasis, oxidative stress, and antioxidant defense. *Front Plant Sci., 13*. https://doi.org/10.3389/fpls.2022.950120.

Ma, J., Saleem, M. H., Yasin, G., Mumtaz, S., Qureshi, F. F., Ali, B., Ercisli, S., Alhag, S. K., Ahmed, A. E., Vodnar, D. C., et al., (2022). Individual and combinatorial effects of SNP and NaHS on morpho-physio-biochemical attributes and phytoextraction of chromium through Cr-stressed spinach (*Spinacia oleracea* L.). *Front Plant Sci., 13*, 973740. https://doi.org/10.3389/fpls.2022.973740.

Martínez-Viveros, O., Jorquera, M. A., Crowley, D. E., Gajardo, G., & Mora, M. L., (2010). Mechanisms and practical considerations involved in plant growth promotion by rhizobacteria. *J. Soil Sci. Plant Nutr., 10*(3), 293–319.

Mayak, S., Tirosh, T., & Glick, B. R., (2004). Plant growth-promoting bacteria confer resistance in tomato plants to salt stress. *Plant Physiol. Biochem., 42*(6), 565–572.

Mehmood, S., Khatoon, Z., Amna, Ahmad, I., Muneer, M. A., Kamran, M. A., Ali, J., Ali, B., et al., (2021). Bacillus sp. PM31 harboring various plant growth-promoting activities regulates *Fusarium* dry rot and wilt tolerance in potato. *Arch. Agron. Soil Sci.,* 1–15. https://doi.org/10.1080/03650340.2021.1971654.

Metayi, M. H., Abd El-Naby, S. S., El-Habal, N. A., Fahmy, H. H., Abdou, M. S., Ali, B., Abdel-Rheim, K. H., & Abdel-Megeed, A., (2022). Omani frankincense nanoemulsion formulation efficacy and its latent effects on biological aspects of the spiny bollworm *Earias insulana* (Boisd.). *Front. Physiol., 13,* 2129. https://doi.org/10.3389/fphys.2022.1001136.

Nawaz, H., Ali, A., Saleem, M. H., Ameer, A., Hafeez, A., Alharbi, K., Ezzat, A., et al., (2022). Comparative effectiveness of EDTA and citric acid assisted phytoremediation of Ni contaminated soil by using canola (*Brassica napus*). *Brazilian J. Biol., 82.* https://doi.org/10.1590/1519-6984.261785.

Naz, R., Khan, M. S., Hafeez, A., Fazil, M., Khan, M. N., Ali, B., Javed, M. A., et al., (2022). Assessment of phytoremediation potential of native plant species naturally growing in a heavy metal-polluted industrial soils. *Brazilian J. Biol., 84.*

Nowak, J., (1998). Benefits of *in vitro* "biotization" of plant tissue cultures with microbial inoculants. *Vitr. Cell Dev. Biol., 34,* 122–130.

Rosenblueth, M., & Martínez-Romero, E., (2006). Bacterial endophytes and their interactions with hosts. *Mol. Plant-Microbe. Interact., 19*(8), 827–837.

Saeed, S., Ullah, A., Ullah, S., Noor, J., Ali, B., Khan, M. N., Hashem, M., et al., (2022). Validating the impact of water potential and temperature on seed germination of wheat (*Triticum aestivum* L.) via hydrothermal time model. *Life, 12*(7). https://doi.org/10.3390/life12070983.

Saini, A., Manuja, S., Kumar, S., Hafeez, A., Ali, B., & Poczai, P., (2022). Impact of cultivation practices and varieties on productivity, profitability, and nutrient uptake of rice (*Oryza sativa* L.) and wheat (*Triticum aestivum* L.) cropping system in India. *Agriculture, 12*(10), 1678.

Salam, A., Afridi, M. S., Javed, M. A., Saleem, A., Hafeez, A., Khan, A. R., Zeeshan, M., Ali, B., Azhar, W., Sumaira, et al., (2022). Nano-priming against abiotic stress: A way forward towards sustainable agriculture. *Sustainability, 14*(22). https://doi.org/10.3390/su142214880.

Saleem, A., Zulfiqar, A., Ali, B., Naseeb, M. A., Almasaudi, A. S., & Harakeh, S., (2022). Iron sulfate ($FeSO_4$) improved physiological attributes and antioxidant capacity by reducing oxidative stress of *Oryza sativa* L. cultivars in alkaline soil. *Sustainability, 14*(24), 16845.

Saleem, K., Asghar, M. A., Saleem, M. H., Raza, A., Kocsy, G., Iqbal, N., Ali, B., et al., (2022). Chrysotile-asbestos-induced damage in panicum virgatum and *Phleum pretense* species and its alleviation by organic-soil amendment. *Sustainability, 14*(17), 10824.

Scholla, M. H., & Elkan, G. H., (1984). *Rhizobium fredii* sp. Nov., a fast-growing species that effectively nodulates soybeans. *Int. J. Syst. Evol. Microbiol., 34*(4), 484–486.

Sekar, S., & Kandavel, D., (2010). Interaction of plant growth promoting rhizobacteria (PGPR) and endophytes with medicinal plants–new avenues for phytochemicals. *J. Phytol., 2*(7).

Solanki, M. K., Solanki, A. C., Rai, S., Srivastava, S., Kashyap, B. K., Divvela, P. K., Kumar, S., Yandigeri, M. S., Kashyap, P. L., Shrivastava, A. K., et al., (2022). Functional interplay between antagonistic bacteria and *Rhizoctonia solani* in the tomato plant rhizosphere. *Front. Microbiol., 13.* https://doi.org/10.3389/fmicb.2022.990850.

Sturz, A. V., Christie, B. R., & Nowak, J., (2000). Bacterial endophytes: Potential role in developing sustainable systems of crop production. *CRC Crit. Rev. Plant Sci., 19*(1), 1–30.

Verma, J. P., Yadav, J., Tiwari, K. N., Lavakush, S., & Singh, V., (2010). Impact of plant growth promoting rhizobacteria on crop production. *Int. J. Agric. Res., 5*(11), 954–983.

Wahab, A., Abdi, G., Saleem, M. H., Ali, B., Ullah, S., Shah, W., Mumtaz, S., et al., (2022). Plants' physio-biochemical and phyto-hormonal responses to alleviate the adverse effects of drought stress: A comprehensive review. *Plants [Internet], 11*(13), 1620. https://doi.org/10.3390/plants11131620.

Wille, L., Messmer, M. M., Studer, B., & Hohmann, P., (2019). Insights to plant–microbe interactions provide opportunities to improve resistance breeding against root diseases in grain legumes. *Plant Cell Environ., 42*(1), 20–40.

Zainab, N., Amna, Khan, A. A., Azeem, M. A., Ali, B., Wang, T., Shi, F., Alghanem, S. M., Munis, M. F. H., Hashem, M., et al., (2021). PGPR-mediated plant growth attributes and metal extraction ability of *Sesbania sesban* L. In industrially contaminated soils. *Agronomy [Internet], 11*(9), 11. https://doi.org/10.3390/agronomy11091820.

Zhang, F., Dashti, N., Hynes, R. K., & Smith, D. L., (1996). Plant growth promoting rhizobacteria and soybean [*Glycine max* (L.) Merr.] nodulation and nitrogen fixation at suboptimal root zone temperatures. *Ann. Bot., 77*(5), 453–460.

Index

O

P

T